Lecture Notes in Mathematics

Edited by A. Dold and B. Eckmann

Series: Université de Nice
Adviser: J. Dieudonné

449

Hyperfunctions and Theoretical Physics

Rencontre de Nice, 21–30 Mai 1973

Edited by F. Pham

Springer-Verlag
Berlin · Heidelberg · New York 1975

Prof. Frédéric Pham
Institut de Mathématiques
et Sciences Physiques
Université de Nice
Parc Valrose
F-06034 Nice-Cedex

Library of Congress Cataloging in Publication Data
Main entry under title:

Hyperfunctions and theoretical physics.

(Lecture notes in mathematics ; 449)
English or French.
Bibliography: p.
Includes index.
1. Mathematical physics--Congresses. 2. Hyperfunctions--Congresses. 3. Quantum field theory--Congresses.
I. Pham, Frédéric. II. Series: Lecture notes in mathematics (Berlin) ; 449.
QA3.L28 no. 449 [QC19.2] 510'.8s [530.1'5] 75-9931

AMS Subject Classifications (1970): 32D10, 46F15, 81A17, 81A48

ISBN 3-540-07151-2 Springer-Verlag Berlin · Heidelberg · New York
ISBN 0-387-07151-2 Springer-Verlag New York · Heidelberg · Berlin

Offsetdruck: Julius Beltz, Hemsbach/Bergstr.

A la mémoire de A. MARTINEAU

Le séjour à NICE, pendant l'année 1972-1973, de M. SATO et de ses élèves T. KAWAI et M. KASHIWARA donna l'occasion à des mathématiciens niçois d'organiser, en liaison avec des physiciens théoriciens, un colloque sur la théorie des hyperfonctions et sur quelques problèmes connexes rencontrés en Physique mathématique.

Au cours de ce colloque, qui a réuni une quarantaine de participants du 21 au 30 Mai 1973, on a pu entendre d'une part des exposés plus ou moins initiatiques (les mathématiciens s'adressant aux physiciens) sur la théorie des hyperfonctions, d'autre part des exposés sur divers problèmes de Théorie Quantique Relativiste en rapport avec la théorie des hyperfonctions : problèmes de théorie des champs d'une part, théorie de la matrice S d'autre part (rappelons, pour les situer rapidement, que ces deux théories s'intéressent aux mêmes phénomènes, à savoir les interactions des particules lourdes à haute énergie, mais que la première essaye de bâtir une axiomatique autour du concept de " champ ", inspiré de l'électrodynamique quantique, tandis que la seconde reste plus " en surface des phénomènes " avec l'entité directement mesurable qu'est " la matrice S ") .

Les textes groupés dans ce volume sont des rédactions d'exposés faits au Colloque. Malheureusement, nous n'avons pas réussi à obtenir de rédaction des exposés de M. SATO, ni de la suite d'exposés de M. KASHIWARA intitulée

" Regularity of hyperfunctions ; applications : Jost points, edge of the wedge; relation between support and singular support " ,

ni de la suite d'exposés de J. BROS intitulée

" Représentation de Fourier locale des hyperfonctions et microfonctions ; théorèmes du type edge of the wedge "

(dont une partie toutefois est résumée par l'Appendice de l'article de D. IAGOLNITZER) .

TABLE DES MATIERES

PART I : HYPERFUNCTIONS

PART II : S-MATRIX

PART III : THEORY OF FIELDS

INTRODUCTION AUX HYPERFONCTIONS

A. CEREZO, J. CHAZARAIN, A. PIRIOU
Département de Mathématiques, NICE

CHAPITRE 0 - INTRODUCTION

Depuis longtemps, on a fait les deux remarques suivantes :

1) Il est utile d'introduire et d'étudier des " fonctions généralisées "

2) Celles-ci ont souvent intérêt à être considérées comme " valeurs au bord " de fonctions très " régulières " (holomorphes, harmoniques,...).

Par exemple, sur $\mathbb{R}$ une distribution T à support compact est valeur au bord de la fonction $\psi(z) = \frac{1}{2i\pi} < T_x, \frac{1}{x-z} >$, holomorphe dans $\mathbb{C}$ - supp T , au sens des distributions :

$$\int_{\mathbb{R}} \left[\psi(x+i\varepsilon) - \psi(x-i\varepsilon)\right] \varphi(x)\, dx \xrightarrow[\varepsilon \to 0^+]{} < T, \varphi > \quad (\forall \varphi \in C_0^\infty(\mathbb{R})) .$$

C'est SATO qui le premier a étudié des fonctions généralisées définies a priori comme " valeurs au bord " de fonctions holomorphes, obtenant ainsi la classe la plus large de fonctions généralisées localisables, les hyperfonctions.

A une variable, les choses sont très simples, car la notion de valeur au bord est immédiate : une hyperfonction sur $\mathbb{R}$ est définie par une fonction holomorphe dans $\mathbb{C} - \mathbb{R}$, et deux telles fonctions définissent la même hyperfonction si et seulement si leur différence est holomorphe dans tout $\mathbb{C}$ (intuitivement, elles ont même saut sur $\mathbb{R}$) .

A plusieurs variables, la notion de valeur au bord est plus délicate, car il faut maîtriser des phénomènes nouveaux de prolongement analytique (par exemple toute fonction holomorphe dans $\mathbb{C}^2 - \mathbb{R}^2$ l'est en fait dans $\mathbb{C}^2$).

SATO a été amené naturellement à rechercher la bonne définition de valeur-au-bord-de-fonction-holomorphe (par exemple sur $\mathbb{R}^n$), dans la cohomologie relative de $\mathbb{R}^n$ dans $\mathbb{C}^n$ à coefficients dans $\mathcal{O}$, c'est-à-dire dans la colonne de gauche de la suite exacte de cohomologie relative :

$$\begin{array}{l} 0 \longrightarrow \mathcal{O}(\mathbb{C}^n) \longrightarrow \mathcal{O}(\mathbb{C}^n - \mathbb{R}^n) \\ \longrightarrow H^1_{\mathbb{R}^n}(\mathbb{C}^n, \mathcal{O}) \longrightarrow H^1(\mathbb{C}^n, \mathcal{O}) \longrightarrow H^1(\mathbb{C}^n - \mathbb{R}^n, \mathcal{O}) \\ \longrightarrow H^2_{\mathbb{R}^n}(\mathbb{C}^n, \mathcal{O}) \longrightarrow \cdots\cdots\cdots \\ \cdots\cdots\cdots \longrightarrow H^{n-1}(\mathbb{C}^n - \mathbb{R}^n, \mathcal{O}) \\ \longrightarrow H^n_{\mathbb{R}^n}(\mathbb{C}^n, \mathcal{O}) \longrightarrow H^n(\mathbb{C}^n, \mathcal{O}) \longrightarrow H^n(\mathbb{C}^n - \mathbb{R}^n, \mathcal{O}) \longrightarrow 0 \quad . \end{array}$$

En fait, il a montré que tous les termes de la première colonne sont nuls sauf le dernier, d'où la définition des hyperfonctions sur $\mathbb{R}^n$,

$$\mathcal{B}(\mathbb{R}^n) = H^n_{\mathbb{R}^n}(\mathbb{C}^n, \mathcal{O})$$

et plus généralement, si Ω est un ouvert de $\mathbb{R}^n$ et U un ouvert de $\mathbb{C}^n$ dans lequel Ω est contenu et fermé :

$$\mathcal{B}(\Omega) = H^n_{\Omega}(U, \mathcal{O}) \quad .$$

Bien entendu, ces fonctions généralisées n'ont pas en général de " valeur " en un point au sens usuel : comme pour les distributions, on ne sait définir que leur restriction à un ouvert. Mais les hyperfonctions sont localisables, en ce sens qu'elles sont déterminées par leurs restrictions aux ouverts d'un recouvrement. Aussi le langage des faisceaux s'impose.

Le faisceau $\mathcal{B}$ des hyperfonctions contient celui des distributions et a l'avantage d'être flasque : toute hyperfonction sur un ouvert peut se prolonger à un ouvert plus grand.

En calculant ces groupes de cohomologie relative à l'aide de la cohomologie de Čech, on fait apparaître naturellement toute hyperfonction comme somme de " valeurs au bord " de fonctions holomorphes dans des tubes du type $\mathbb{R}^n + i\Gamma$, où Γ est un cône convexe ouvert propre de $\mathbb{R}^n$. Cette représentation permet de décomposer la singularité d'une hyperfonction suivant les directions cotangentes à $\mathbb{R}^n$, et SATO construit ainsi le faisceau $\mathcal{C}$ des microfonctions, sur le fibré en sphères cotangentes. Le support de la microfonction associée à une hyperfonction n'est autre que son spectre singulier (wave front).

Toutes ces constructions sont naturelles, et beaucoup d'opérations définies sur les fonctions holomorphes se prolongent donc naturellement aux hyperfonctions et aux microfonctions.

SOMMAIRE .

CHAPITRE I . HYPERFONCTIONS A 1 VARIABLE (A. CEREZO)

Il concerne le cas de la dimension 1 , sur lequel on ne s'étend que pour introduire des notions utiles aux autres chapitres. Il contient aussi ce qu'il faut de théorie des faisceaux pour en utiliser le langage.

CHAPITRE II . HYPERFONCTIONS A UN NOMBRE QUELCONQUE DE VARIABLES (A. PIRIOU)

On donne la définition des hyperfonctions au moyen des groupes de cohomologie relative, et on montre que toute hyperfonction est une somme finie de valeurs au bord de fonctions holomorphes dans des tubes locaux.

CHAPITRE III . FAISCEAU $\mathcal{C}$. (J. CHAZARAIN)

On expose la notion de spectre singulier d'une hyperfonction et on définit le faisceau $\mathcal{C}$ des singularités.

CHAPITRE IV . APPLICATIONS . (J. CHAZARAIN)

On donne un énoncé hyperfonction du théorème " edge of the wedge " puis on décrit les opérations que l'on peut définir dans certains cas sur les hyperfonctions.

Références . Le propos de ces exposés est purement " pédagogique ", aussi on n'y trouvera rien de nouveau au sujet des hyperfonctions. On a utilisé de nombreuses sources, à commencer par les publications et les exposés de SATO et de ses deux élèves KASHIWARA et KAWAI. On s'est également inspiré des présentations de la théorie par KOMATSU ainsi que par MORIMOTO. Enfin, pour le lien entre distributions et hyperfonctions, on renvoie aux travaux de MARTINEAU et au livre de SHAPIRA.

On trouvera ces références dans :

M. SATO . Theory of hyperfunctions I and II - J. Fac. of Sciences Univ. Tokyo (1959) & (1960)

M. SATO . Regularity of hyperfunction solutions of partial differential equations - Proc. NICE Congress, 2 , Gauthiers-Villars, Paris (1970)

Les articles de SATO-KAWAI-KASHIWARA, H. KOMATSU, M. MORIMOTO dans Hyperfunctions and pseudo-differential equations , Proc. Conf. Katata, Lecture Note n° 287, Springer (1973)

A. MARTINEAU . Distributions et valeurs au bord des fonctions holomorphes - Proc. Inter. Summer Inst. Lisbon (1964)

P. SHAPIRA . Théorie des hyperfonctions - Lecture Note N° 126, Springer (1970)

CHAPITRE I - LE CAS D'UNE VARIABLE

A. CEREZO

I DEFINITION

La suite exacte de cohomologie relative se réduit dans le cas d'une seule variable à :

$$0 \longrightarrow \mathcal{O}(\mathbb{C}) \longrightarrow \mathcal{O}(\mathbb{C}-\mathbb{R}) \longrightarrow H^1_{\mathbb{R}}(\mathcal{O},\mathbb{C}) \longrightarrow 0$$

et on définit donc les hyperfonctions sur $\mathbb{R}$ par

$$\mathcal{B}(\mathbb{R}) = \frac{\mathcal{O}(\mathbb{C}-\mathbb{R})}{\mathcal{O}(\mathbb{C})} = (= H^1_{\mathbb{R}}(\mathbb{C},\mathcal{O}))$$

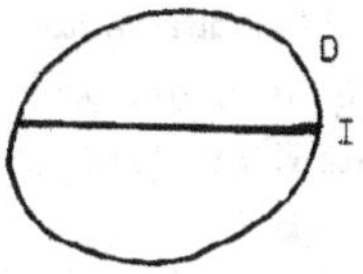

et plus généralement, si I est un ouvert de $\mathbb{R}$, et D un voisinage complexe de I (c'est-à-dire un ouvert de $\mathbb{C}$ dans lequel I est contenu et fermé),

$$\mathcal{B}(I) = \frac{\mathcal{O}(D-I)}{\mathcal{O}(D)} \quad (= H^1_I(D,\mathcal{O})) \quad .$$

Une hyperfonction sur I est donc la donnée d'une fonction holomorphe dans $D-I$, où D est un certain voisinage complexe de I, modulo les fonctions qui se prolongent en fonctions holomorphes dans tout D. Intuitivement, une hyperfonction sur I, c'est le saut sur I d'une fonction holomorphe auprès de I.

Le théorème d'excision affirme que $\mathcal{B}(I)$ est indépendant du voisinage complexe D choisi : si D_1 et D_2 sont deux voisinages complexes de I, on a un isomorphisme canonique

$$\frac{\mathcal{O}(D_1-I)}{\mathcal{O}(D_1)} \xrightarrow{\sim} \frac{\mathcal{O}(D_2-I)}{\mathcal{O}(D_2)}$$

isomorphisme qui, lorsque $D_1 \supset D_2$, est induit par la restriction naturelle $\mathcal{O}(D_1-I) \longrightarrow \mathcal{O}(D_2-I)$.

<u>Remarque</u> - On définirait de façon analogue les hyperfonctions sur une variété analytique réelle de dimension 1 .

Il est facile de restreindre une hyperfonction à un ouvert plus petit $I' \subset I$, par restriction d'une fonction représentative : si $I' \subset I$ et $D' - I' \subset D - I$, la restriction $\mathcal{O}(D-I) \longrightarrow \mathcal{O}(D'-I')$ induit une application $\mathcal{B}(I) \longrightarrow \mathcal{B}(I')$, appelée <u>restriction</u> (et qu'on note comme telle). On vérifie immédiatement que ces restrictions sont transitives : si $I'' \subset I' \subset I$ et $f \in \mathcal{B}(I)$,

$$f|_{I'}|_{I''} = f|_{I''} \quad .$$

On a ainsi défini un <u>préfaisceau</u> $\mathcal{B}$ sur $\mathbb{R}$, qui se trouve être un <u>faisceau</u>.

II - <u>QUELQUES NOTIONS UTILES SUR LES FAISCEAUX</u>

La donnée, pour tout ouvert U d'un espace topologique X , d'un espace vectoriel (sur $\mathbb{C}$) $\mathcal{F}(U)$ appelé espace des <u>sections au-dessus de</u> U , et, pour tout couple d'ouverts (U,U') tels que $U' \subset U$, d'une application linéaire appelée <u>restriction</u> de $\mathcal{F}(U)$ dans $\mathcal{F}(U')$, de telle sorte que

- la restriction de U dans U soit l'identité
- si $U'' \subset U' \subset U$ le diagramme

$$\begin{array}{ccc} \mathcal{F}(U) & \longrightarrow & \mathcal{F}(U'') \\ & \searrow \quad \nearrow & \\ & \mathcal{F}(U') & \end{array}$$

est commutatif (id est : $\forall f \in \mathcal{F}(U) \quad f|_{U''} = f|_{U'}|_{U''}$)

constitue ce qu'on appelle un <u>préfaisceau</u> (d'espaces vectoriels sur $\mathbb{C}$) <u>$\mathcal{F}$ sur la base X</u> .

Un tel préfaisceau est dit être un <u>faisceau</u> si " on peut recoller les sections locales de manière unique ", c'est-à-dire si :

Pour tout recouvrement d'un ouvert U par une famille d'ouverts $(U_\alpha)_{\alpha \in A}$, et pour toute donnée de sections $f_\alpha \in \mathcal{F}(U_\alpha)$ " compatibles " (c'est-à-dire telles que $f_\alpha|_{U_\alpha \cap U_\beta} = f_\beta|_{U_\alpha \cap U_\beta}$ pour tous α, β dans A) , il existe une et une seule section $f \in \mathcal{F}(U)$ telle que $f|_{U_\alpha} = f_\alpha$ pour tout α .

Exemples (sur $\mathbb{R}^n$) — Le faisceau des fonctions continues, celui des fonctions indéfiniment dérivables, celui des distributions

- le faisceau des fonctions quelconques (à valeurs complexes)
- le faisceau $\mathfrak{A}$ des fonctions analytiques, le faisceau $\mathcal{O}$ (sur $\mathbb{C}^n$) des fonctions holomorphes
- le préfaisceau des fonctions bornées n'est pas un faisceau.

THEOREME – 1) Le préfaisceau $\mathcal{B}$ des hyperfonctions sur $\mathbb{R}$ est un faisceau

2) Ce faisceau est flasque (id est : toutes les restrictions sont surjectives).

On admettra ici le premier point qui équivaut essentiellement au lemme de Cousin. Le second est facile : toute hyperfonction f sur I' peut se représenter par une fonction φ holomorphe dans $[D - (I-I')] - I'$, puisque le crochet est un voisinage complexe de I' ; φ est holomorphe dans D-I et définit donc une hyperfonction sur I dont la restriction à I' est évidemment f .

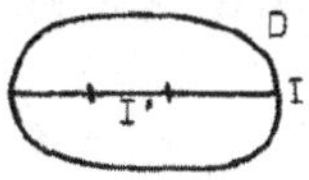

Remarque – De tous les exemples cités ci-dessus, seul le faisceau des fonctions quelconques est flasque. On sait en particulier qu'une distribution sur un ouvert ne se prolonge pas en général à un ouvert plus grand.

Quand on a un préfaisceau qui n'est pas un faisceau, on sait construire un faisceau associé qui lui est attaché de manière naturelle : au-dessus de chaque ouvert, on prend comme sections " toutes les sections locales qui se recollent " : si (U_α) est un recouvrement de U , on considère l'espace vectoriel $\mathcal{F}((U_\alpha))$ des familles (f_α) de sections au-dessus des ouverts du recouvrement, qui se recollent :

$$f_\alpha \in \mathcal{F}(U_\alpha) \quad \text{et} \quad f_\alpha|_{U_\alpha \cap U_\beta} = f_\beta|_{U_\alpha \cap U_\beta} .$$

On construit $\mathcal{F}((U_\alpha))$ pour tous les recouvrements (U_α) de U , et dans la réunion des $\mathcal{F}((U_\alpha))$, on identifie deux familles qui coïncident localement (c'est-à-dire par restriction aux ouverts d'un recouvrement encore plus fin). Le résultat c'est

$$\mathcal{L}\mathcal{F}(U) = \varinjlim \mathcal{F}((U_\alpha))$$

où la limite inductive est prise sur tous les recouvrements de U . $\mathcal{L}\mathcal{F}(U)$ est alors l'espace des sections au-dessus de U d'un préfaisceau $\mathcal{L}\mathcal{F}$.

Si $\mathcal{F}$ était un préfaisceau séparé (c'est-à-dire si toute section localement nulle de $\mathcal{F}$ est nulle), il se trouve que $\mathcal{L}\mathcal{F}$ est un faisceau, et c'est lui qu'on appelle le faisceau associé à $\mathcal{F}$. (Sinon, il faut recommencer, et considérer $\mathcal{L}\mathcal{L}\mathcal{F}$, mais peu importe ici, car tous les préfaisceaux qu'on considèrera dans la suite seront séparés).

On peut résumer cette construction en disant qu'intuitivement, les sections du faisceau associé sont les gens qui sont localement des sections du préfaisceau (au moins si ce dernier est séparé).

Exemple - Le faisceau associé au préfaisceau des fonctions bornées (sur $\mathbb{R}^n$) est le faisceau des fonctions localement bornées (c'est-à-dire bornées sur tout compact).

Un morphisme de préfaisceaux $\mathcal{F} \longrightarrow \mathcal{G}$, où $\mathcal{F}$ et $\mathcal{G}$ sont deux préfaisceaux sur la même base X, c'est la donnée pour tout ouvert U de X d'une application linéaire de $\mathcal{F}(U)$ dans $\mathcal{G}(U)$ de telle sorte que ces applications commutent aux restrictions : si $U' \subset U$ le diagramme

$$\begin{array}{ccc} \mathcal{F}(U) & \longrightarrow & \mathcal{G}(U) \\ \downarrow & & \downarrow \\ \mathcal{F}(U') & \longrightarrow & \mathcal{G}(U') \end{array} \qquad \text{est commutatif.}$$

La notion de faisceau associé à un préfaisceau est naturelle en ce sens que :

1) Il existe un morphisme canonique $\mathcal{F} \longrightarrow \mathcal{L}\mathcal{F}$, qui n'est autre que l'identité si $\mathcal{F}$ est un faisceau.

2) Un morphisme de préfaisceaux $\mathcal{F} \longrightarrow \mathcal{G}$ se prolonge en un morphisme des faisceaux associés $\mathcal{L}\mathcal{F} \longrightarrow \mathcal{L}\mathcal{G}$ et le diagramme

$$\begin{array}{ccc} \mathcal{F} & \longrightarrow & \mathcal{G} \\ \downarrow & & \downarrow \\ \mathcal{L}\mathcal{F} & \longrightarrow & \mathcal{L}\mathcal{G} \end{array} \qquad \text{est commutatif}$$

En particulier, si $\mathcal{G}$ est un faisceau, tout morphisme $\mathcal{F} \longrightarrow \mathcal{G}$ se prolonge en un morphisme $\mathcal{L}\mathcal{F} \longrightarrow \mathcal{G}$ ...

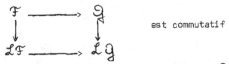

Remarque - On peut parler du faisceau associé à un préfaisceau dont on ne s'est donné les sections qu'au-dessus de certains ouverts, formant une base d'ouverts de X : on pourra en effet trouver un recouvrement plus fin qu'un recouvrement donné de X et qui soit formé d'ouverts de la base, et il suffit donc de considérer les familles de sections au-dessus de ces ouverts là qui se recollent.

Le faisceau en question est, si l'on veut, le faisceau associé au préfaisceau obtenu en choisissant $\{0\}$ comme espace de sections au-dessus des autres ouverts.

Cet abus de langage est bien pratique, et sera commis systématiquement dans la suite.

III - PREMIERS CALCULS SUR LES HYPERFONCTIONS

Puisque $\mathcal{B}(I) = \dfrac{\mathcal{O}(D-I)}{\mathcal{O}(D)}$, toute hyperfonction $f(x)$ sur I se représente par une fonction $\varphi(z)$ holomorphe dans $D-I$, ce qu'on écrira :
$f(x) = [\varphi(z)]_{z=x}$.

On sait multiplier une hyperfonction par une fonction analytique : si $a(x) \in \mathcal{Q}(I)$, elle se prolonge en fonction holomorphe dans un voisinage complexe D' de I. $D \cap D'$ est encore un voisinage complexe de I, et on pose

$$a(x)\, f(x) = [a(z)\, \varphi(z)]_{z=x} .$$

On sait aussi dériver une hyperfonction :

$$\frac{df}{dx}(x) = \left[\frac{d\varphi}{dz}(z)\right]_{z=x} .$$

Plus généralement, si $P(x, \frac{d}{dx})$ est un opérateur différentiel linéaire à coefficients analytiques sur I, on posera

$$P(x, \frac{d}{dx})\, f(x) = \left[P(z, \frac{d}{dz})\, \varphi(z)\right]_{z=x}$$

où la fonction entre crochets est holomorphe dans un certain voisinage complexe de I (l'intersection de D et de voisinages où les coefficients de P se prolongent en fonctions holomorphes).

Injections des fonctions analytiques :

On pose $\varepsilon(z) = \begin{cases} 1 & \text{si } \operatorname{Im} z > 0 \\ 0 & \text{si } \operatorname{Im} z < 0 \end{cases}$, $\quad \overline{\varepsilon}(z) = \begin{cases} 0 & \text{si } \operatorname{Im} z > 0 \\ -1 & \text{si } \operatorname{Im} z < 0 \end{cases}$

et $1(x) = [\varepsilon(z)]_{z=x}$ ($=[\bar{\varepsilon}(z)]_{z=x}$ puisque $\varepsilon - \bar{\varepsilon}$ est analytique partout).

On a alors une injection des fonctions analytiques sur I dans les hyperfonctions sur I (pour tout I, $\mathcal{A}(I) \to \mathcal{B}(I)$) :

$$a(x) \in \mathcal{A}(I) \longmapsto a(x).1(x) = [a(z)\,\varepsilon(z)]_{z=x} \in \mathcal{B}(I)$$

c'est-à-dire qu'on identifie la fonction $a(x)$ analytique sur I avec l'hyperfonction définie par

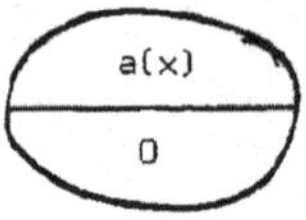

I ou

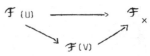

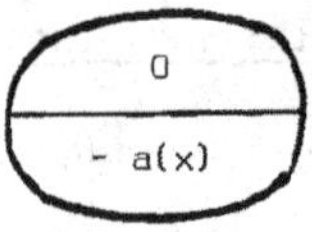

Il s'agit d'une injection de faisceaux $\mathcal{A} \longrightarrow \mathcal{B}$.

IV - MORPHISMES DE FAISCEAUX

On sait restreindre à un ouvert plus petit une section d'un préfaisceau (par exemple une distribution ou une hyperfonction), mais on ne sait pas a priori définir sa valeur en un point. On peut tout au plus lui donner un sens de la manière suivante : si $x \in U$ et $f \in \mathcal{F}(U)$, on regarde les restrictions de f à des ouverts de plus en plus petits contenant x . On identifie f et g si elles coïncident dans un petit voisinage de x . Autrement dit, on considère l'image de f dans

$$\mathcal{F}_x = \lim_{V \ni x} \mathcal{F}(V)$$

et on appelle cette image <u>germe</u> de f en x . Dès que $x \in V \subset U$, on a un diagramme commutatif

$$\begin{array}{ccc} \mathcal{F}(U) & \longrightarrow & \mathcal{F}_x \\ & \searrow \quad \nearrow & \\ & \mathcal{F}(V) & \end{array}$$

L'espace $\mathcal{F}_x$ s'appelle la <u>fibre</u> de $\mathcal{F}$ en x , et ses éléments sont les <u>germes de sections</u> de $\mathcal{F}$ en x . Quand on a un morphisme de préfaisceaux $\mathcal{F} \to \mathcal{G}$ (c'est-à-dire pour tout U , $\mathcal{F}(U) \to \mathcal{G}(U)$ ...) on en déduit naturellement des morphismes entre les fibres : pour tout x , $\mathcal{F}_x \to \mathcal{G}_x$

avec des diagrammes commutatifs

$$\begin{array}{ccc} \mathcal{F}(U) & \longrightarrow & \mathcal{G}(U) \\ \downarrow & & \downarrow \\ \mathcal{F}_x & \longrightarrow & \mathcal{G}_x \end{array} \qquad \text{dès que } x \in U$$

On dit qu'un morphisme de faisceaux $\mathcal{F} \to \mathcal{G}$ est une injection, ou une surjection, si et seulement si c'est vrai sur chaque fibre (pour tout x $\mathcal{F}_x \longrightarrow \mathcal{G}_x$ est injectif, surjectif). On pourra parler aussi d'une suite exacte de faisceaux $\longrightarrow \mathcal{F} \xrightarrow{u} \mathcal{G} \xrightarrow{v} h \longrightarrow$: ça veut dire qu'en chaque point x la suite

$$\longrightarrow \mathcal{F}_x \xrightarrow{u} \mathcal{G}_x \xrightarrow{v} h_x \longrightarrow$$

est exacte (c'est-à-dire $\mathrm{Im}\, u = \mathrm{Ker}\, v$) : $g \in \mathcal{G}_x$ vérifie $v(g) = 0$ si et seulement s'il existe $f \in \mathcal{F}_x$ telle que $u(f) = g$.

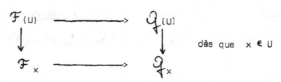

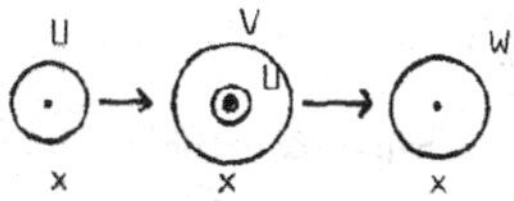

En termes de sections, g se représente par une section de $\mathcal{G}$ dans un voisinage V de x , et l'exactitude en x signifie que l'image par v de cette section est nulle localement (pas forcément dans V mais au moins dans un voisinage W de x peut-être plus petit) si et seulement si elle est dans l'image d'une section de $\mathcal{F}$ au-dessus d'un voisinage U de x localement (c'est-à-dire qu'elles peuvent ne coïncider ni dans V ni dans U , mais peut être seulement dans un voisinage encore plus petit de x).

On notera $0 \longrightarrow \mathcal{F} \longrightarrow \mathcal{G}$ une injection et $\mathcal{F} \longrightarrow \mathcal{G} \longrightarrow 0$ une surjection, parce que les suites écrites sont exactes si 0 signifie le faisceau nul (dont tous les espaces de sections, et donc toutes les fibres sont $\{0\}$) .

Par exemple, la suite $0 \longrightarrow \mathcal{A} \longrightarrow \mathcal{B}$ est une injection de faisceaux sur $\mathbb{R}$, qu'on complètera bientôt en une suite excate

$$0 \longrightarrow \mathcal{A} \longrightarrow \mathcal{B} \longrightarrow \pi_* \mathcal{C} \longrightarrow 0 \ .$$

V - AUTRES CALCULS

Donnons maintenant un sens précis à la notion de Valeurs au bord :

Si $\varphi \in \mathcal{O}(D-I)$, on pose $\varphi(x + io) = [\varepsilon(z)\ \varphi(z)]$

$\varphi(x - io) = [\bar{\varepsilon}(z)\ \varphi(z)]$

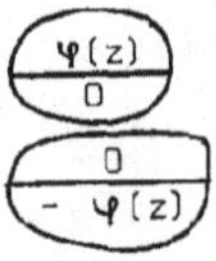

et on appelle l'hyperfonction $\varphi(x + io)$ $(\varphi(x - io))$ la valeur au bord par dessus (par dessous) de φ .

Avec ces notations ,

$$[\varphi(z)]_{z=x} = \varphi(x + io) - \varphi(x - io) = \text{" saut de } \varphi \text{ "} .$$

Ces notations concrétisent l'idée intuitive que l'hyperfonction représentée par φ n'est pas autre chose que le saut de φ , c'est-à-dire la différence de ses valeurs au bord par dessus et par dessous.

Support ; intégration des hyperfonctions à support compact :

Comme $\mathcal{B}$ est un faisceau, on a la notion de support d'une hyperfonction : c'est le complémentaire du plus grand ouvert (ou de la réunion des ouverts) où sa restriction est nulle.

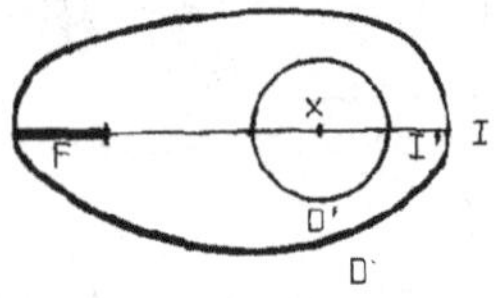

Si $f = [\varphi] \in \mathcal{B}(I)$ a son support dans un fermé F et si $x \in I-F$, il existe $I' \subset I$ voisinage de x tel que $f|_{I'} = 0$.

Si $D' \subset D$ est un voisinage complexe de I' tel que $D' - I' \subset D - I$, $f|_{I'}$ est définie par $\varphi|_{D'-I'}$ Donc φ est en fait holomorphe dans D' tout entier.

En bref :

$$\operatorname{supp} f \subset F \iff \varphi \in \mathcal{O}(D-F) .$$

Soit maintenant K un compact de I et $f = [\varphi]$ une hyperfonction sur I à support dans K ce qui se note $f \in \mathcal{B}_K(I)$. On définit l'intégrale de f sur I par la formule

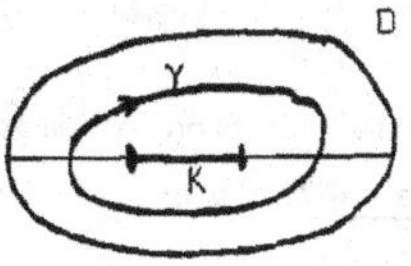

$$\int_I f(x)\,dx = \int_\gamma \varphi(z)\,dz$$

où γ est un contour tracé dans $D - K$ et entourant K une fois dans le sens inhabituel. Il faut remarquer que grâce au théorème de Cauchy la définition ne dépend ni du choix du représentant φ, ni du choix du contour γ. Cette définition devient intuitive si l'on resserre γ autour de K et qu'on se souvient que f n'est autre que le saut de φ.

Exemples d'hyperfonctions :

Si $\alpha(z) \in \mathcal{O}(D)$, la formule de Cauchy affirme que

$$\frac{-1}{2i\pi} \int_\gamma \frac{\alpha(z)}{z-z_o}\,dz = \alpha(z_o)$$

lorsque γ est un contour entourant le compact $\{z_o\}$ comme ci-dessus. On est donc conduit à définir l'hyperfonction de Dirac par

$$\delta(x) = \frac{-1}{2i\pi}\left[\frac{1}{z}\right],$$

de sorte que l'on a :

$$\delta \in \mathcal{B}_{\{0\}}(\mathbb{R}), \qquad \int_{\mathbb{R}} \delta(x)\,dx = 1 \text{ , et plus généralement}$$

$$\int_I \alpha(x)\cdot\delta(x)\,dx = \alpha(0) \quad \text{dès que} \quad \alpha \in \mathcal{Q}(I) \text{ , } I \text{ contenant l'origine.}$$

On remarquera que dans la notation par valeurs au bord, on a :

$$\delta(x) = \frac{-1}{2i\pi}\left(\frac{1}{x+io} - \frac{1}{x-io}\right)$$

On obtient en dérivant $\delta^{(n)}(x) = \dfrac{(-1)^{n+1}\,n!}{2i\pi}\left[\dfrac{1}{z^{n+1}}\right]$

et en intégrant l'hyperfonction de Heaviside :

$$H(x) = \frac{-1}{2i\pi}\left[\log(-z)\right]$$

où log est la détermination principale, holomorphe sauf sur la demi-droite réelle négative.

On se contentera d'un seul autre exemple : si f est une fonction méromorphe sur I, on définit une hyperfonction " partie finie de f " par

$$\mathrm{Pf}\, f(x) = \frac{1}{2}\left[f(x+i0) + f(x-i0)\right]$$

($f(x+i0) - f(x-i0)$ serait une hyperfonction portée par les pôles, c'est-à-dire une somme de combinaisons linéaires finies de dérivées de δ en ces points).

En particulier $\mathrm{Pf}\,\frac{1}{x} = \frac{1}{2}\left(\frac{1}{x+i0} + \frac{1}{x-i0}\right)$

d'où la formule bien connue

$$\frac{1}{x\pm i0} = \mathrm{Pf}\,\frac{1}{x} \mp i\pi\delta(x)$$

Changement de variables .

En fait, on peut sous certaines conditions, définir beaucoup plus généralement l'image réciproque ou l'image directe d'une hyperfonction par une application analytique, et ce sera fait dans le Chapitre IV . On se contente ici du cas particulier d'un difféomorphisme : si $\xi : I_1 \to I_2$ est un difféomorphisme analytique de I_1 sur I_2 (c'est-à-dire si ξ' ne s'annule pas), ξ se prolonge en une bijection holomorphe d'un voisinage D_1 de I_1 sur un voisinage D_2 de I_2 .

Si alors $g = [\psi(z)]_{z=x} \in \mathcal{B}(I_2)$, où $\psi \in \mathcal{O}(D_2 - I_2)$, on définit l'image réciproque f de g par ξ par la formule :

$$f(x) = \xi^* g(x) = (\text{signe de } \xi')\,[\psi(\xi(z))]_{z=x} ,$$

qui généralise l'image réciproque d'une fonction analytique.

Plongement des distributions :

Si K est un compact de I , et $f \in \mathcal{B}_K(I) \simeq \mathcal{B}_K(\mathbb{R})$, on obtient une fonction représentative φ_0 de f holomorphe dans $\mathbb{C} - K$ en posant :

$$\varphi_0(z) = \frac{1}{2i\pi}\int_{\mathbb{R}} \frac{f(x)}{x-z}\,dx \qquad (\text{on montre que } [\varphi_0] = f) .$$

C'est aussi la manière dont on plonge les distributions dans les hyperfonctions : si $T \in \mathcal{E}'(\mathbb{R})$, on pose :

$$\varphi(z) = \frac{1}{2i\pi} < T_x, \frac{1}{x-z} >$$

et l'application $T \longmapsto [\varphi]$ ainsi obtenue est une injection de $\mathcal{E}'(R)$ dans $\mathcal{B}(R)$ qui conserve le support. On peut toujours décomposer une distribution sur $\mathbb{R}$ en somme localement finie de distributions à support compact, identifier celles-ci à des hyperfonctions par la méthode précédente, puis sommer dans $\mathcal{B}$ la famille ainsi obtenue, qui est à supports localement en nombre fini, puisque $\mathcal{B}$ est un faisceau. On obtient ainsi une injection de faisceaux $0 \longrightarrow \mathcal{D}' \longrightarrow \mathcal{B}$.

Convolution -

On se contente ici de signaler qu'on peut définir le produit de convolution de deux hyperfonctions sur $\mathbb{R}$ sous les mêmes conditions de support que pour des distributions, et que ce produit jouit de propriétés (commutativité, support,...) en tous points semblables.

VI - SPECTRE SINGULIER D'UNE HYPERFONCTION

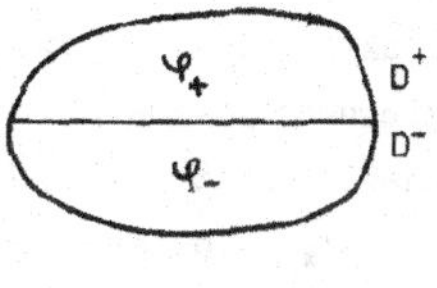

On utilise la notation

$$f = [\varphi] = \varphi_+(x+i0) - \varphi_-(x-i0)$$

où $f \in \mathcal{B}(I)$, φ en est un représentant holomorphe dans $D-I$, $D-I$ est la réunion de deux morceaux D^+ et D^- contenus l'un dans le demi-plan supérieur, l'autre dans le demi-plan inférieur (ce qui suppose qu'on les a distingués), et $\varphi_\pm = \varphi|_{D^\pm}$.

DEFINITIONS - On dit que f est analytique en x_0 $(\in I)$ si φ_+ et φ_- se prolongent en fonctions holomorphes au voisinage de x_0 .

- On dit que f est microanalytique en $x_0 + i\infty$ $(x_0 - i\infty)$ φ_+ (φ_-) se prolonge en fonction holomorphe au voisinage de x_0 .

Remarques . 1) Ces propriétés ne dépendent pas du représentant φ de f .

2) La première définition est cohérente avec l'injection $\mathcal{A} \longrightarrow \mathcal{B}$ déjà définie : si f est analytique en x_o , il existe un voisinage I_o de x_o et un voisinage complexe D_o de I_o tels que φ_+ et φ_- sont holomorphes dans D_o , donc $f(x) = \varphi_+(x+io) - \varphi_-(x-io) = \varphi_+(x+io) - \varphi_-(x+io)$

$$= [(\varphi_+(z) - \varphi_-(z))\varepsilon(z)] \text{ dans } I_o ,$$

et c'est dire que $f|_{I_o}$ est dans l'image de $\mathcal{A}(I_o) \longrightarrow \mathcal{B}(I_o)$.

3) Dire que f est microanalytique en x+i∞ (x-i∞) , c'est dire qu'au voisinage de x_o , f est une valeur au bord par-dessous (par-dessus) : en effet, une restriction convenable de f s'écrit

$$\varphi_+(x-io) - \varphi_-(x-io) \qquad (\varphi_+(x+io) - \varphi_-(x+io)) \quad .$$

Pour rendre compte de la microanalyticité, on est amené à considérer deux exemplaires de I , notés I+i∞ et I-i∞ ,et on notera S^*I leur réunion disjointe (I+i∞) ⊔ (I-i∞) . (Dans le cas des hyperfonctions de plusieurs variables sur un ouvert Ω , c'est en effet le fibré en sphères cotangentes $S^*\Omega$ que l'on introduit pour étudier la microanalyticité ; ici les espaces tangent et cotangent en un point de I sont des droites et leurs sphères sont donc réduites chacune à deux points).

Le spectre singulier de f , noté SSf est, par définition, le complémentaire dans S^*I de l'ensemble des points où f est microanalytique : un point x±i∞ de I±i∞ est dans le spectre singulier de f si et seulement si $\varphi_\pm$ ne se prolonge pas au voisinage de x .

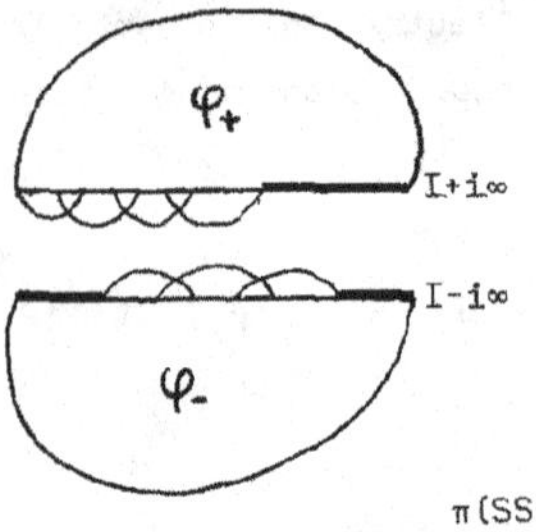

SSf est évidemment un fermé de S^*I , et si $\pi : S^*I \longrightarrow I$ est la projection canonique (c'est-à-dire l'application $x \pm i\infty \overset{\pi}{\longmapsto} x$) , on a visiblement :

$$\pi(SSf) = \text{supp sing } f \quad ,$$

où supp sing f est le support singulier (analytique) de f , c'est-à-dire le complémentaire dans I de l'ensemble des points où f est analytique.

<u>Remarque</u> . Le spectre singulier d'une hyperfonction porte aussi quelquefois dans la littérature les noms de <u>support essentiel</u> (noté SE) , ou de <u>Wave front</u> (noté WF) , ou même de support singulier, bien que ce dernier terme crée une confusion regrettable avec le supp sing défini ci-dessus, qui en est la projection sur la base.

<u>Exemples</u> : $SS\ \frac{1}{x+i0} = \{0+i\infty\}$ $\quad$ $\text{supp sing}\ \frac{1}{x+i0} = \{0\}$ $\quad$ $\text{supp}\ \frac{1}{x+i0} = \mathbb{R}$

$SS\ \delta = \{0+i\infty, 0-i\infty\}$ $\quad$ $\text{supp sing}\ \delta = \{0\}$ $\quad$ $\text{supp}\ \delta = \{0\}$

<u>Applications</u> .

On en verra plusieurs dans les chapitres suivants. On se contente ici d'une seule, à titre d'exemple.

Les hyperfonctions sont des êtres trop singuliers pour qu'on sache les multiplier entre elles en général. Mais on le peut sous certaines conditions qui portent sur leur spectre singulier.

Soit a l'application <u>antipodale</u> de $S^* I$, qui échange les deux exemplaires de I : $a(x \pm i\infty) = x \mp i\infty$.

<u>THEOREME</u> . <u>Si</u> f <u>et</u> g <u>sont deux hyperfonctions sur</u> I , <u>et sous l'hypothèse</u>

$$SS\ f \cap a(SS\ g) = \emptyset ,$$

<u>le produit</u> $f.g \in \mathcal{B}(I)$ <u>est défini naturellement</u> . (En particulier, ce produit prolonge celui des fonctions analytiques, ou d'une fonction analytique par une hyperfonction).

Puisque $\mathcal{B}$ est un faisceau, il suffit en effet de définir ce produit localement puis de recoller. Or localement, l'hypothèse du théorème implique que l'on a l'une des quatre situations suivantes (les hâchures figurent la non-microanalyticité d'un côté) :

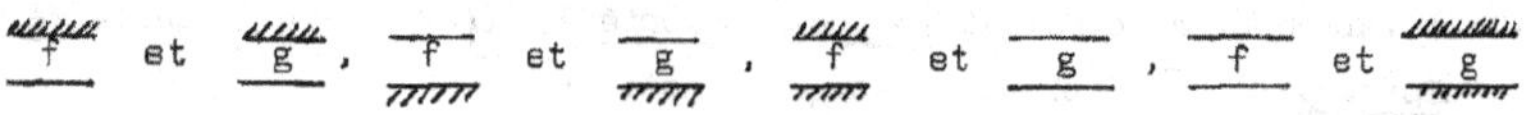
f
et
g
f
et
g
f
et
g
f
et
g

Dans les deux derniers cas, l'une des deux hyperfonctions est analytique, et le produit a déjà été défini. Dans les deux premiers cas, f et g sont des valeurs au bord du même côté ; par exemple dans le premier cas, on a $f(x) = \varphi(x+i0)$ et $g(x) = \psi(x+i0)$. Il suffit alors de poser

$$f.g(x) = \varphi\psi(x+i0) .$$

Par exemple une expression comme δ_+^2 est bien définie : c'est (à un coefficient près) l'hyperfonction

$$\left(\frac{1}{x+i0}\right)^2 = \left[\frac{1}{z}\,\varepsilon(z)\right]^2 = \left[\frac{1}{z^2}\,\varepsilon(z)\right]$$

VII - IMAGES DE FAISCEAUX

On n'a défini de morphismes de faisceaux qu'entre faisceaux sur une même base. Il est nécessaire de savoir transporter un faisceau d'une base sur une autre par une application continue. Cela se fait dans les deux sens.

Faisceau image directe :

Soient X et Y deux espaces topologiques, $X \xrightarrow{u} Y$ une application continue, et $\mathcal{F}$ un faisceau sur X. On définit un préfaisceau $u_* \mathcal{F}$ sur Y en posant $u_* \mathcal{F}(V) = \mathcal{F}(u^{-1}(V))$ pour tout ouvert V de Y. On vérifie immédiatement que $u_* \mathcal{F}$ est un faisceau, qu'on appelle le faisceau image directe de $\mathcal{F}$ par u.

Faisceau image inverse :

Soient X, Y, u comme précédemment et $\mathcal{F}$ un faisceau sur Y. La construction du faisceau inverse se complique du fait que l'image par u d'un ouvert de X n'est pas un ouvert. On obtient quand même un préfaisceau $\mathcal{G}$ sur X en posant, pour tout ouvert U de X.

$$\mathcal{G}(U) = \varinjlim_{V \supset u(U)} \mathcal{F}(V)$$

(Au lieu de prendre les sections de $\mathcal{F}$ sur $u(U)$ qui n'est pas ouvert, on a pris toutes les sections de $\mathcal{F}$ sur tous les voisinages V de $u(U)$, en identifiant celles qui coïncident au voisinage de $u(U)$. Si $u(U)$ est ouvert, on retrouve bien $\mathcal{G}(U) = \mathcal{F}(u(U))$). Mais le préfaisceau $\mathcal{G}$ n'est pas en général un faisceau, et c'est le faisceau associé à $\mathcal{G}$ qu'on appelle le faisceau image réciproque de $\mathcal{F}$ par u, et qu'on note $u^{-1}\mathcal{F}$.

Ses sections sont donc compliquées à décrire en général, mais on se consolera en remarquant que ses fibres, elles, s'expriment simplement ; on a en effet :

$$(u^{-1}\mathcal{F})_x = \mathcal{F}_{u(x)} .$$

VIII - LE FAISCEAU $\mathcal{C}$ DES MICROFONCTIONS

On introduit ici de nouvelles notations dont l'utilité se fera surtout sentir dans le cas de plusieurs variables (chapitres suivants). Il s'agit de construire plusieurs faisceaux sur la base $S^*\mathbb{R} = (\mathbb{R}+i\infty) \cup (\mathbb{R}-i\infty)$.

- D'abord, comme l'application $\pi : S^*\mathbb{R} \to \mathbb{R}$ est ici particulièrement simple, il est aisé de construire le faisceau $\pi^{-1}\mathcal{B}$ sur $S^*\mathbb{R}$: les sections de $\pi^{-1}\mathcal{B}$ au-dessus d'un ouvert de la forme $I\pm i\infty$ (c'est-à-dire contenu dans $\mathbb{R}+i\infty$ ou dans $\mathbb{R}-i\infty$) sont tout simplement les hyperfonctions sur I . Bien entendu, sur un ouvert de la forme $(I+i\infty) \cup (J-i\infty)$, l'espace des sections de $\pi^{-1}\mathcal{B}$ est la somme directe $\mathcal{B}(I) \oplus \mathcal{B}(J)$.

- On construit ensuite un sous-faisceau $\mathcal{A}^*$ du précédent en considérant le préfaisceau sur $S^*\mathbb{R}$ dont les sections sur un ouvert de la forme $I+i\infty$ (ou $I-i\infty$) sont les hyperfonctions sur I qui sont microanalytiques en tous les points de $I+i\infty$ (ou $I-i\infty$) . Le faisceau associé $\mathcal{A}^*$ a les mêmes sections au dessus des ouverts de cette forme, et pour sections au dessus d'un ouvert quelconque la somme directe des sections au dessus de chaque composante. On a visiblement une injection de faisceaux sur $S^*\mathbb{R}$:

$$0 \longrightarrow \mathcal{A}^* \longrightarrow \pi^{-1}\mathcal{B} .$$

- Enfin le faisceau $\mathcal{C}$ (dont les sections sont appelées microfonctions) est le faisceau-quotient de $\pi^{-1}\mathcal{B}$ par $\mathcal{A}^*$. On construit d'abord le préfaisceau quotient, dont les sections au-dessus de l'ouvert U sont le quotient

$$\frac{\pi^{-1}\mathcal{B}(U)}{\mathcal{A}^*(U)}$$

Soit ici, si $U = (I+i\infty) \cup (J-i\infty)$,

$$\frac{\pi^{-1}\mathcal{B}(U)}{\mathcal{A}^*(U)} = \frac{\mathcal{B}(I) \oplus \mathcal{B}(J)}{\mathcal{A}^*(I+i\infty) \oplus \mathcal{A}^*(J-i\infty)} = \frac{\mathcal{B}(I)}{\mathcal{A}^*(I+i\infty)} \oplus \frac{\mathcal{B}(J)}{\mathcal{A}^*(J-i\infty)} .$$

En général, quand on construit ainsi le préfaisceau quotient d'un faisceau par un sous-faisceau, on n'obtient pas un faisceau, et ce sera le cas, par exemple, quand on fera la construction analogue pour les hyperfonctions de plusieurs variables. C'est alors le faisceau associé qu'on appelle faisceau quotient.

Ici, par contre, on obtient tout de suite un faisceau, le faisceau $\mathcal{C}$ des microfonctions.

Par exemple, l'espace $\mathcal{C}(I+i\infty)$, des microfonctions sur $I+i\infty$, est l'espace des hyperfonctions sur I considérées modulo celles qui sont microanalytiques sur $I+i\infty$ (ou encore modulo celles qui sont valeurs au bord par dessous).

Il résulte de la construction même du faisceau $\mathcal{C}$ qu'on a une suite exacte de faisceaux sur $S^* I$:

$$0 \longrightarrow \mathcal{A}^* \longrightarrow \pi^{-1}\mathcal{B} \longrightarrow \mathcal{C} \longrightarrow 0$$

qui d'ailleurs peut se transporter ici en une suite exacte de faisceaux sur I :

$$0 \longrightarrow \mathcal{A} \longrightarrow \mathcal{B} \xrightarrow{sp} \pi_* \mathcal{C} \longrightarrow 0 \quad .$$

Les quatre faisceaux $\mathcal{B}$, $\pi^{-1}\mathcal{B}$, $\mathcal{C}$, $\pi_*\mathcal{C}$ sont flasques, mais pas $\mathcal{A}^*$ (ni $\mathcal{A}$!).

L'application " spectre ", notée sp, de $\mathcal{B}$ dans $\pi_*\mathcal{C}$, fait correspondre à une hyperfonction sa " singularité " qui est une microfonction : en effet, si $f \in \mathcal{B}(I)$,

$$sp\ f \in \pi_* \mathcal{C}(I) = \mathcal{C}(\pi^{-1}(I)) = \mathcal{C}((I+i\infty) \cup (I-i\infty)) \quad .$$

Enfin, il est clair par construction qu'on a l'égalité suivante, pour une hyperfonction f :

$$SS\ f = \text{supp}\ (sp\ f)$$

le second membre est le support de la microfonction $sp\ f$, en tant que section du faisceau $\mathcal{C}$.

On achève ce chapitre en énonçant une proposition (locale) et son corollaire immédiat (global) ; la première n'est qu'une autre manière de prononcer l'égalité ci-dessus.

PROPOSITION - Une hyperfonction f est valeur au bord par dessus (par dessous) au voisinage d'un point x, si et seulement si

$$SS\ f \not\ni x - i\infty \qquad (x+i\infty)$$

COROLLAIRE - Une hyperfonction f sur I est de la forme $\varphi(x \pm io)$ si et seulement si

$$SS\ f \subset I \pm i\infty \quad .$$

CHAPITRE II - HYPERFONCTIONS A UN NOMBRE QUELCONQUE DE VARIABLES

A. PIRIOU

I - INTRODUCTION HEURISTIQUE

On a vu dans l'exposé précédent qu'une hyperfonction sur $\mathbb{R}$ est un élément de $\mathcal{B}(\mathbb{R}) = \dfrac{\mathcal{O}(\mathbb{C} - \mathbb{R})}{\mathcal{O}(\mathbb{C})}$ - soient $u_1, u_2 \in \mathcal{B}(\mathbb{R})$, représentées respectivement par $\varphi_1, \varphi_2 \in \mathcal{O}(\mathbb{C} - \mathbb{R})$; il est naturel de chercher à définir l'hyperfonction (de deux variables) $u(x_1,x_2) = u_1(x_1)\, u_2(x_2)$ au moyen de la fonction $\varphi(z_1,z_2) = \varphi_1(z_1)\, \varphi_2(z_2)$; remarquons que φ est holomorphe dans $(\mathbb{C} - \mathbb{R}) \times (\mathbb{C} - \mathbb{R})$, et qu'elle est définie modulo $\mathcal{O}(\mathbb{C} \times (\mathbb{C} - \mathbb{R})) + \mathcal{O}((\mathbb{C} - \mathbb{R}) \times \mathbb{C})$ puisque φ_1 et φ_2 ne sont définies que modulo $\mathcal{O}(\mathbb{C})$ - Ainsi, on est conduit à définir u comme élément de

$$\frac{\mathcal{O}((\mathbb{C} - \mathbb{R}) \times (\mathbb{C} - \mathbb{R}))}{\mathcal{O}(\mathbb{C} \times (\mathbb{C} - \mathbb{R})) + \mathcal{O}((\mathbb{C} - \mathbb{R}) \times \mathbb{C})} \qquad \text{qui n'est autre,}$$

comme on le verra plus loin, que $H^2(\mathbb{C}^2 \bmod \mathbb{C}^2 - \mathbb{R}^2, \mathcal{O})$ second groupe de cohomologie relative de $\mathbb{C}^2$ modulo $\mathbb{C}^2 - \mathbb{R}^2$ à valeurs dans le faisceau $\mathcal{O}$, et qu'on note aussi $H^2_{\mathbb{R}^2}(\mathbb{C}^2, \mathcal{O})$.

D'autre part, si on reprend l'interprétation d'une hyperfonction sur $\mathbb{R}$ en termes de valeurs au bord abstraites on a :

$$u_1(x_1) = \varphi_1(x_1 + io) - \varphi_1(x_1 - io)$$
$$u_2(x_2) = \varphi_2(x_2 + io) - \varphi_2(x_2 - io)$$

et il est naturel d'écrire $u(x_1,x_2) = u_1(x_1)\, u_2(x_2)$ sous forme d'une somme de quatre "valeurs au bord" :

$$u(x_1,x_2) = \varphi(x_1 + io,\ x_2 + io) - \varphi(x_1 - io,\ x_2 + io)$$
$$+ \varphi(x_1 - io,\ x_2 - io) - \varphi(x_1 + io,\ x_2 - io)$$

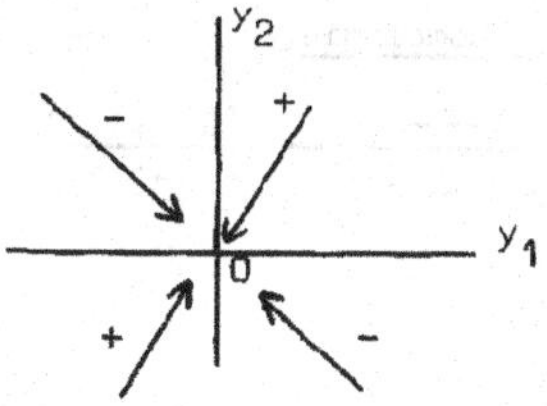

Dans ce qui suit, nous allons définir les hyperfonctions de n variables au moyen de la cohomologie relative, et préciser la notion de valeur au bord, qui nous fournira un moyen commode de représenter les hyperfonctions. Pour cela, il nous faut tout d'abord rappeler brièvement la cohomologie de Čech. On désignera par $\mathcal{O}$ le faisceau (sur $\mathbb{C}^n$) des fonctions holomorphes.

II - RESUME DE LA COHOMOLOGIE DE ČECH

Soient U, U' deux ouverts de $\mathbb{C}^n$, avec $U' \subset U$. Appelons recouvrement relatif $(\mathcal{U},\mathcal{U}')$ de (U,U') la donnée d'un recouvrement ouvert $\mathcal{U} = (U_\alpha)_{\alpha \in I}$ de U, et d'un sous-recouvrement $\mathcal{U}' = (U_\alpha)_{\alpha \in I'}$, (avec $I' \subset I$) de U'. Pour $p \in \mathbb{N}$, le groupe additif $C^p(\mathcal{U},\mathcal{U}',\mathcal{O})$ des p - cochaines alternées (relatives) de $(\mathcal{U},\mathcal{U}')$ à coefficients dans $\mathcal{O}$ est constitué par les familles $\varphi = (\varphi_{\alpha_0,\dots,\alpha_p})_{(\alpha_0,\dots,\alpha_p) \in I^{p+1}}$ telles que

$$\begin{cases} a) & \varphi_{\alpha_0,\dots,\alpha_p} \in \mathcal{O}(U_{\alpha_0\dots\alpha_p}), \text{ où on a posé } U_{\alpha_0\dots\alpha_p} = U_{\alpha_0} \cap \dots \cap U_{\alpha_p}. \\ b) & \varphi_{\alpha_0,\dots,\alpha_i,\dots,\alpha_j,\dots,\alpha_p} = -\varphi_{\alpha_0,\dots,\alpha_j,\dots,\alpha_i,\dots,\alpha_p} \\ c) & \varphi_{\alpha_0,\dots,\alpha_p} = 0 \text{ si } \alpha_0,\dots,\alpha_p \in I' \end{cases}$$

On définit ensuite les morphismes de cobord δ

$$\delta : C^p(\mathcal{U},\mathcal{U}',\mathcal{O}) \longrightarrow C^{p+1}(\mathcal{U},\mathcal{U}',\mathcal{O})$$
$$\varphi \longmapsto \delta\varphi \qquad \text{par :}$$

$$(\delta\varphi)_{\alpha_0,\dots,\alpha_{p+1}} = \sum_{j=0}^{p+1} (-1)^j \varphi_{\alpha_0,\dots,\hat{\alpha}_j,\dots,\alpha_{p+1}}$$

On vérifie facilement que $\delta \circ \delta = 0$, d'où un complexe de groupes :

$$0 \longrightarrow C^0(\mathcal{U},\mathcal{U}',\mathcal{O}) \xrightarrow{\delta} C^1(\mathcal{U},\mathcal{U}',\mathcal{O}) \xrightarrow{\delta} \dots \xrightarrow{\delta} C^{p-1}(\mathcal{U},\mathcal{U}',\mathcal{O}) \xrightarrow{\delta}$$
$$C^p(\mathcal{U},\mathcal{U}',\mathcal{O}) \xrightarrow{\delta} C^{p+1}(\mathcal{U},\mathcal{U}',\mathcal{O}) \longrightarrow \dots$$

dont on considère les groupes de cohomologie

$$H^p(\mathcal{U},\mathcal{U}',\mathcal{O}) = \frac{\operatorname{Ker}\,(C^p(\mathcal{U},\mathcal{U}',\mathcal{O}) \xrightarrow{\delta} C^{p+1}(\mathcal{U},\mathcal{U}',\mathcal{O}))}{\operatorname{Im}\,(C^{p-1}(\mathcal{U},\mathcal{U}',\mathcal{O}) \xrightarrow{\delta} C^p(\mathcal{U},\mathcal{U}',\mathcal{O}))}.$$

Soit maintenant $(\mathcal{V},\mathcal{V}')$ un autre recouvrement relatif de (U,U'), avec $\mathcal{V} = (V_\beta)_{\beta \in J}$, $\mathcal{V}' = (V_\beta)_{\beta \in J'}$ $(J' \subset J)$.
Nous dirons que le recouvrement $(\mathcal{V}, \mathcal{V}')$ est <u>plus fin</u> que le recouvrement $(\mathcal{U},\mathcal{U}')$ si :

a) Pour tout $\beta \in J$, il existe $\alpha \in I$ tel que $V_\beta \subset U_\alpha$

b) Pour tout $\beta \in J'$, il existe $\alpha \in I'$ tel que $V_\beta \subset U_\alpha$

Soit alors une fonction de choix $\alpha = \tau(\beta)$; définissons les morphismes

$$C^p(\mathcal{U},\mathcal{U}',\mathcal{O}) \xrightarrow{\tau^*} C^p(\mathcal{V},\mathcal{V}',\mathcal{O})$$

$$\varphi \longmapsto \tau^*\varphi \quad \text{par :}$$

$(\tau^*\varphi)_{\beta_0,\ldots,\beta_p} = \varphi_{\tau(\beta_0),\ldots,\tau(\beta_p)}$. Les morphismes τ^*, qui commutent évidemment avec les morphismes de cobord δ, induisent des morphismes $H^p(\mathcal{U},\mathcal{U}',\mathcal{O}) \longrightarrow H^p(\mathcal{V},\mathcal{V}',\mathcal{O})$ qui sont <u>canoniques</u>, en ce sens qu'ils ne dépendent plus de la fonction de choix τ utilisée.

<u>DEFINITION 1</u> - On appelle p^{ieme} groupe de cohomologie relative de U modulo U' à valeurs dans le faisceau $\mathcal{O}$ le groupe $H^p(U \bmod U', \mathcal{O}) = \varinjlim_{(\mathcal{U},\mathcal{U}')} H^p(\mathcal{U},\mathcal{U}',\mathcal{O})$, où la limite inductive est prise selon les recouvrements plus fins.

Signalons que ce groupe est aussi noté $H^p_{U-U'}(U,\mathcal{O})$; dans le cas où $U' = \emptyset$, on le note simplement $H^p(U,\mathcal{O})$.

Précisons que, dans la définition 1, le passage à la <u>limite inductive</u> signifie ceci : soient $(\mathcal{U},\mathcal{U}')$ et $(\mathcal{V},\mathcal{V}')$ deux recouvrements relatifs de (U,U'), alors $\varphi \in H^p(\mathcal{U},\mathcal{U}',\mathcal{O})$ et $\Psi \in H^p(\mathcal{V},\mathcal{V}',\mathcal{O})$ sont <u>identifiés</u> s'il existe un recouvrement relatif $(\mathfrak{w},\mathfrak{w}')$ de (U,U'), plus fin que les deux précédents, tel que φ et Ψ aient la même image par les morphismes canoniques

$$\begin{array}{ccc} H^p(\mathcal{U},\mathcal{U}',\mathcal{O}) & \searrow & \\ & & H^p(\mathfrak{w},\mathfrak{w}',\mathcal{O}) \\ H^p(\mathcal{V},\mathcal{V}',\mathcal{O}) & \nearrow & \end{array}$$

Remarquons que si $(\mathcal{U},\mathcal{U}')$ est un recouvrement relatif quelconque de (U,U'), on a des morphismes canoniques $H^p(\mathcal{U},\mathcal{U}',\mathcal{O}) \longrightarrow H^p(U,U',\mathcal{O})$.

Mais si on suppose $H^q(U_{\alpha_0,\ldots,\alpha_p}, \mathcal{O}) = 0$ pour tout $q \geq 1$, pour tout $p \geq 0$, et pour tous $\alpha_0, \ldots, \alpha_p \in I$, on sait alors (<u>théorème de Leray</u>) que ces morphismes sont en fait des <u>isomorphismes</u>. Ces conditions sont en particulier vérifiées si, pour tout $\alpha \in I$, U_α est un <u>ouvert d'holomorphie</u> de $\mathbb{C}^n$. On a donc le

<u>THEOREME 1</u> - Soit $(\mathcal{U}, \mathcal{U}')$ un recouvrement relatif de (U, U'), avec $\mathcal{U} = (U_\alpha)_{\alpha \in I}$ et U_α ouvert d'holomorphie pour tout $\alpha \in I$

Alors $H^p(U \bmod U', \mathcal{O}) \simeq H^p(\mathcal{U}, \mathcal{U}', \mathcal{O})$.

III - <u>DEFINITION DES HYPERFONCTIONS</u>

Soit Ω un ouvert de $\mathbb{R}^n$, et considérons un <u>voisinage complexe</u> U de Ω, c'est-à-dire un ouvert U de $\mathbb{C}^n$ tel que Ω soit contenu et fermé dans U. Alors $U' = U - \Omega$ est ouvert dans U, et on peut considérer les groupes de cohomologie relative:

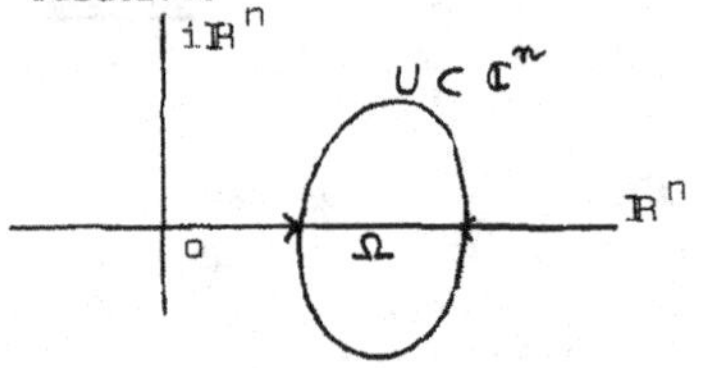

$$H^p(U \bmod U - \Omega, \mathcal{O}) = H^p_\Omega(U, \mathcal{O})$$

Soit maintenant ω ouvert dans Ω, et V un voisinage complexe de ω tel que $V \subset U$, $V - \omega \subset U - \Omega$; si $(\mathcal{U}, \mathcal{U}')$ est un recouvrement relatif de $(U, U-\Omega)$, on obtient un recouvrement relatif $(\mathcal{V}, \mathcal{V}')$ de $(V, V-\omega)$ en posant $V_\alpha = U_\alpha \cap V$; par restriction, on obtient des morphismes $C^p(\mathcal{U}, \mathcal{U}', \mathcal{O}) \longrightarrow C^p(\mathcal{V}, \mathcal{V}', \mathcal{O})$, qui induisent des morphismes $H^p(\mathcal{U}, \mathcal{U}', \mathcal{O}) \longrightarrow H^p(\mathcal{V}, \mathcal{V}', \mathcal{O})$, d'où finalement des morphismes $\rho^p_{\Omega,\omega} : H^p(U \bmod U-\Omega, \mathcal{O}) \longrightarrow H^p(V \bmod V-\omega, \mathcal{O})$.

Dans le cas où $\omega = \Omega$, on montre (<u>théorème d'excision</u>) que $\rho^p_{\Omega,\omega}$ est un isomorphisme, et donc que $H^p(U \bmod U-\Omega, \mathcal{O})$ ne dépend que de p et de Ω, et non pas du voisinage complexe U utilisé pour Ω ; pour p fixé, on peut maintenant considérer le <u>préfaisceau</u> qui, à tout ouvert Ω de $\mathbb{R}^n$, associe l'espace vectoriel $H^p(U \bmod U-\Omega, \mathcal{O})$, les morphismes de restriction étant les morphismes $\rho^p_{\Omega,\omega}$ définis plus haut. On a alors le résultat fondamental suivant :

THEOREME et DEFINITION 2

1) H^p (U mod U − Ω, $\mathcal{O}$) = 0 si $p \neq n$, Ω ouvert de $\mathbb{R}^n$.

2) Le préfaisceau défini par la donnée, pour tout ouvert Ω de $\mathbb{R}^n$, de l'espace H^n (U mod U − Ω, $\mathcal{O}$), et par les morphismes de restriction $\rho^n_{\Omega,\omega}$ est un faisceau flasque qu'on appelle faisceau $\mathcal{B}$ (sur $\mathbb{R}^n$) des hyperfonctions.

Donc si Ω est un ouvert de $\mathbb{R}^n$, les hyperfonctions sur Ω sont les éléments de l'espace.

$$\mathcal{B}(\Omega) = H^n\ (U \bmod U - \Omega,\ \mathcal{O}) = H^n_\Omega\ (U,\ \mathcal{O})$$

où U est un voisinage complexe arbitraire de Ω.

Dans tout ce qui précède, on peut remplacer $\mathbb{R}^n$ par une variété analytique réelle M de dimension n , et $\mathbb{C}^n$ par une complexifiée X de M.

IV - REPRESENTATIONS D'UNE HYPERFONCTION PAR DES FONCTIONS HOLOMORPHES

Soit Ω un ouvert de $\mathbb{R}^n$; considérons un voisinage complexe U de Ω tel que

$$\begin{cases} a) & U \text{ d'holomorphie} \\ b) & U \cap \mathbb{R}^n = \Omega \end{cases}$$

(rappelons à ce propos que, d'après le théorème de Grauert, tout ouvert de $\mathbb{R}^n$ admet un système fondamental de voisinages d'holomorphie dans $\mathbb{C}^n$).

On va expliciter $\mathcal{B}(\Omega) = H^n$ (U mod U − Ω, $\mathcal{O}$) en utilisant le théorème 1.

Premier exemple de recouvrement :

On prend le recouvrement relatif $(\mathcal{U},\mathcal{U}')$ de (U,U−Ω) défini par :

$$\mathcal{U} = (U_0,\ U_1, \ldots,\ U_n)$$

$$\mathcal{U}' = (U_1, \ldots,\ U_n)$$

$$U_0 = U$$

$$U_j = U \cap \ \{z = (z_1, \ldots, z_n) \in \mathbb{C}^n \mid \operatorname{Im} z_j \neq 0\} \quad (j = 1, \ldots, n)$$

En appliquant les définitions du § 2, on voit que :

$$C^{n+1}(\mathcal{U},\mathcal{U}',\mathcal{O}) = \{o\}$$
$$C^{n}(\mathcal{U},\mathcal{U}',\mathcal{O}) \simeq \mathcal{O}(U \cap (\mathbb{C} - \mathbb{R})^n)$$
$$C^{n-1}(\mathcal{U},\mathcal{U}',\mathcal{O}) \simeq \left\{ \varphi = (\varphi_j)_{j=1,\dots,n} \mid \varphi_j \in \mathcal{O}(U_{1..\hat{j}..n}) \right\}$$

et que $\delta : C^{n-1}(\mathcal{U},\mathcal{U}',\mathcal{O}) \longrightarrow C^{n}(\mathcal{U},\mathcal{U}',\mathcal{O})$ est alors défini par

$$\varphi \longmapsto \delta\varphi = \sum_{j=1}^{n} (-1)^j \varphi_j \ .$$

Puisque $U_o, U_1, \dots, U_n$ sont des ouverts d'holomorphie, les théorèmes 1 et 2 donnent

$$(*) \qquad \boxed{\mathcal{B}(\Omega) = \frac{\mathcal{O}(U \cap (\mathbb{C} - \mathbb{R})^n)}{\sum_{j=1}^{n} \mathcal{O}(U_{1..\hat{j}..n})}}$$

$$\left(\text{avec } U_{1..\hat{j}..n} = \{z \in U \mid \operatorname{Im} z_k \neq 0 \text{ pour } k \neq j\} \right)$$

et l'on retrouve ainsi l'espace-quotient introduit au § 1 dans le cas $\Omega = \mathbb{R}^2$, $U = \mathbb{C}^2$.

Désignons les 2^n composantes connexes de $(\mathbb{C} - \mathbb{R})^n$ par $\mathbb{R}^n + i\Gamma_\sigma$, où $\sigma = (\sigma_1,\dots,\sigma_n)$, avec $\sigma_j = \pm 1$, et où Γ_σ est le cône convexe ouvert $\Gamma_\sigma = \{y \in \mathbb{R}^n \mid \sigma_j y_j > 0 \text{ pour } j = 1,\dots,n\}$. La formule (*) s'écrit

$$(*') \qquad \boxed{\mathcal{B}(\Omega) = \frac{\bigoplus_{\sigma} \mathcal{O}(U \cap (\mathbb{R}^n + i\Gamma_\sigma))}{\sum_{j=1}^{n} \mathcal{O}(U_{1..\hat{j}..n})}}$$

de sorte qu'une hyperfonction $f \in \mathcal{B}(\Omega)$ peut être représentée par 2^n fonctions

$$\varphi_\sigma \in \mathcal{O}(U \cap (\mathbb{R}^n + i\Gamma_\sigma))$$

Deuxième exemple de recouvrement

Soient n + 1 vecteurs $\eta_{1,..,}\ \eta_{n+1}$ de $\mathbb{R}^n - \{o\}$ tels que les n + 1 demi-espaces ouverts

$$\eta_j^{\perp} = \{ y \in \mathbb{R}^n \mid y \,.\, \eta_j > 0 \}$$

recouvrent $\mathbb{R}^n - \{o\}$, c'est-à-dire tels que 0 soit intérieur à l'enveloppe convexe de $\{\eta_{1,\dots,}\eta_{n+1}\}$.

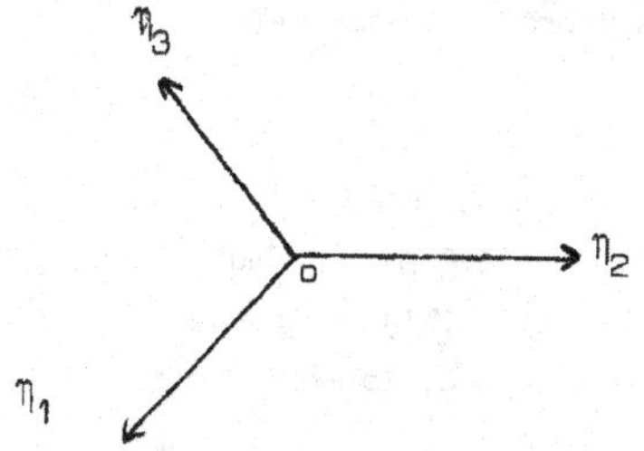

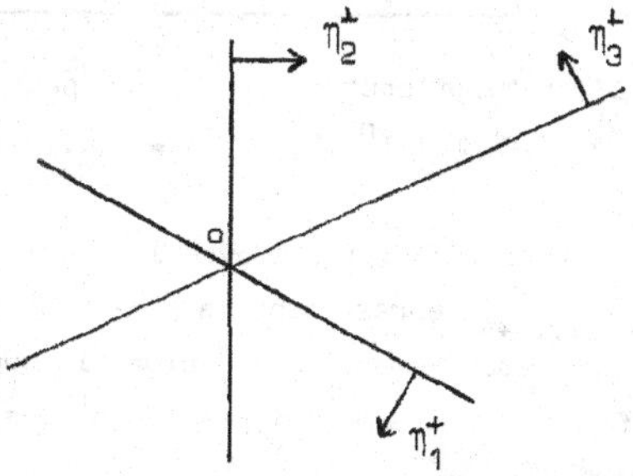

On considère alors le recouvrement relatif $(\mathcal{V},\mathcal{V}')$ de $(U, U - \Omega)$ défini par :

$$\mathcal{V} = (V_o,\ V_{1,\dots,}\ V_{n+1})$$
$$\mathcal{V}' = \quad (V_{1,\dots,}\ V_{n+1})$$
$$V_o = \ V$$
$$V_j = \ U \cap (\mathbb{R}^n + i\eta_j^{\perp}) \quad (j=1,\dots,n+1)$$

On a :

$$C^{n+1}\ (\mathcal{V},\mathcal{V}',\mathcal{O}) \ = \{o\} \ \text{puisque} \ \eta_1^{\perp} \cap\dots\cap \eta_{n+1}^{\perp} = \emptyset$$
$$C^{n}\ (\mathcal{V},\mathcal{V}',\mathcal{O}) \ \simeq \bigoplus_{j=1}^{n+1} \mathcal{O}\,(V_{1..\hat{j}..n+1})$$
$$C^{n-1}\ (\mathcal{V},\mathcal{V}',\mathcal{O}) \ \simeq \bigoplus_{1\le i<j\le n+1} \mathcal{O}(V_{1..\hat{i}..\hat{j}..n+1})$$

et $\delta : C^{n-1}\ (\mathcal{V},\mathcal{V}',\mathcal{O}) \longrightarrow C^n\ (\mathcal{V},\mathcal{V}',\mathcal{O})$ est défini par

$$(\delta\Psi)_j = \sum_{1\le i<j} (-1)^i\ \Psi_{i,j} + \sum_{j<i\le n+1} (-1)^{i-1}\ \Psi_{j,i}\ .$$

Puisque $V_o,\ V_{1,\dots,}\ V_{n+1}$ sont des ouverts d'holomorphie, on obtient, en posant $V_{1..\hat{j}..n+1} = U \cap (\mathbb{R}^n + i\Gamma_j)$, et en désignant par Γ_j le cône ouvert convexe

$$\Gamma_j = \eta_1^{\perp} \cap..\widehat{\cap\eta_j^{\perp}} \cap..\eta_{n+1}^{\perp} \quad ,$$

l'égalité

$$\mathcal{B}(\Omega) = \frac{\bigoplus_{j=1}^{n+1} \mathcal{O}(U\cap(\mathbb{R}^n + i\Gamma_j))}{\sum_{1\le i<j\le n+1} \mathcal{O}(V_{1..\hat{i}..\hat{j}..n+1})} \qquad (**)$$

de sorte qu'une hyperfonction $f \in \mathcal{B}(\Omega)$ peut être représentée par n+1 fonctions $\Psi_j \in \mathcal{O}(U\cap(\mathbb{R}^n+i\Gamma_j))$ $(j = 1,\dots,n+1)$.

Remarque 1 : Pour vérifier si, avec les notations ci-dessus, $\varphi = (\varphi_\sigma)_\sigma$ et $\Psi = (\Psi_j)_{j=1,\dots,n+1}$ représentent la même hyperfonction $f \in \mathcal{B}(\Omega)$, il faut considérer un recouvrement $(\mathcal{W},\mathcal{W}')$ plus fin que $(\mathcal{U},\mathcal{U}')$ et $(\mathcal{V},\mathcal{V}')$, puis voir si φ et Ψ ont, par les morphismes canoniques décrits au § 2, la même image dans $H^n(\mathcal{W},\mathcal{W}',\mathcal{O})$.

Exemple emprunté à A. CEREZO : on prend $\Omega = \mathbb{R}^2$, $U = \mathbb{C}^2$, et on désigne par $f = -4\pi\ \delta(x_1, x_2)$ l'hyperfonction représentée, au moyen de (*), par la fonction $\varphi(z_1,z_2) = \frac{1}{z_1 z_2} \in \mathcal{O}((\mathbb{C}-\mathbb{R})^2)$.

On définit maintenant le recouvrement $(\mathcal{V},\mathcal{V}')$ par
$V_0 = \mathbb{C}^2$, $V_1 = \{z \in \mathbb{C}^2 \mid \mathcal{I}m z_1 > 0\}$, $V_2 = \{z \in \mathbb{C}^2 \mid \mathcal{I}m z_2 > 0\}$,
$V_3 = \{z \in \mathbb{C}^2 \mid \mathcal{I}m z_1 + \mathcal{I}m z_2 < 0\}$, et on cherche à représenter f par (Ψ_1, Ψ_2, Ψ_3), avec $\Psi_j \in \mathcal{O}(\mathbb{R}^n + i\Gamma_j)$ $(j = 1,2,3)$

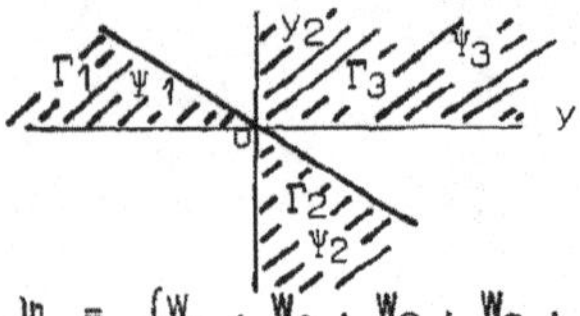

Comme recouvrement $(\mathcal{W},\mathcal{W}')$ plus fin que $(\mathcal{U},\mathcal{U}')$ et $(\mathcal{V},\mathcal{V}')$, prenons par exemple

$\mathcal{W} = (W_0, W_1, W_2, W_3, W_4)$
$\mathcal{W}' = (W_1, W_2, W_3, W_4)$, avec
$W_0 = \mathbb{C}^2$, $W_1 = V_1$, $W_2 = V_2$, $W_3 = \{z \in \mathbb{C} \mid \mathcal{I}m z_1 < 0,\ \mathcal{I}m z_1 + \mathcal{I}m z_2 < 0\}$,
$W_4 = \{z \in \mathbb{C}^2 \mid \mathcal{I}m z_2 < 0,\ \mathcal{I}m z_1 + \mathcal{I}m z_2 < 0\}$

Les inclusions

$$W_o \subset U_o \;,\; V_o$$
$$W_1 \subset U_1 \;,\; V_1$$
$$W_2 \subset U_2 \;,\; V_2$$
$$W_3 \subset U_1 \;,\; V_3$$
$$W_4 \subset U_2 \;,\; V_3$$

permettent d'expliciter les morphismes canoniques

$$\begin{array}{ccc} H^2\,(\mathcal{U},\mathcal{U}',\mathcal{O}) & \searrow & \\ & & H^2\,(\mathfrak{w},\mathfrak{w}',\mathcal{O}) \\ H^2\,(\mathcal{V},\mathcal{V}',\mathcal{O}) & \nearrow & \end{array}$$

et on obtient que (Ψ_1,Ψ_2,Ψ_3) représente f si et seulement si il existe des fonctions $\theta_j \in \mathcal{O}\,(W_j)$ $(j=1,2,3,4)$ telles que :

$$\frac{1}{z_1 z_2} - \Psi_3 = \theta_2 - \theta_1$$
$$- \Psi_2 = \theta_3 - \theta_1$$
$$\frac{1}{z_1 z_2} - \Psi_2 = \theta_4 - \theta_1$$
$$- \frac{1}{z_1 z_2} - \Psi_1 = \theta_3 - \theta_2$$
$$- \Psi_1 = \theta_4 - \theta_2$$
$$\frac{1}{z_1 z_2} = \theta_4 - \theta_3$$

La dernière équation, du type "lemme de Cousin" , admet par exemple la solutio

$$\begin{cases} \theta_4 = \dfrac{1}{z_2(z_1+z_2)} \\ \theta_3 = \dfrac{1}{z_1(z_1+z_2)} \end{cases}$$

Si on choisit par exemple $\theta_1 = o$, $\theta_2 = o$, on obtient finalement que $f = -4\pi\,\delta(x_1,x_2)$ est représentée par

$$\Psi_1 = \frac{-1}{z_1 z_2} - \theta_3 = \frac{-1}{z_2(z_1+z_2)} \,,\quad \Psi_2 = \frac{1}{z_1 z_2} - \theta_4 = \frac{1}{z_1(z_1+z_2)} \quad,\quad \Psi_3 = \frac{1}{z_1 z_2}$$

V - VALEURS AU BORD

On suppose désormais que $\mathbb{R}^n$ est orienté . Soient Ω un ouvert de $\mathbb{R}^n$ et Γ un cône ouvert convexe (de sommet o) de $\mathbb{R}^n$. Soit U un voisinage complexe d'holomorphie de Ω , avec $U\cap\mathbb{R}^n = \Omega$ et soit φ une fonction holomorphe dans le tube local $U\cap(\mathbb{R}^n+i\Gamma)$; on va associer à φ une valeur au bord selon Γ , qui sera une hyperfonction sur Ω. Pour cela, on reprend le deuxième exemple de recouvrement exposé au paragraphe précédent, en choisissant les vecteurs $\eta_1, \eta_2, .., \eta_{n+1}$ tels que

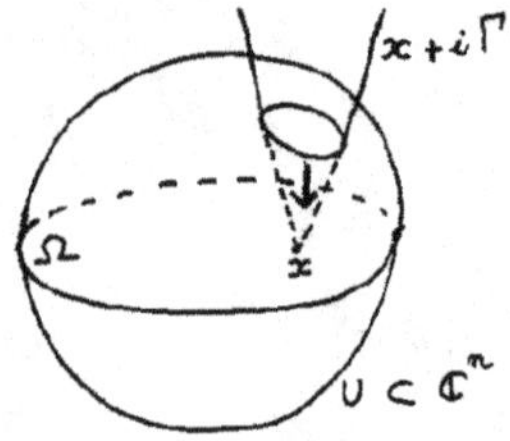

$$\Gamma_1 = \eta_2^\perp \cap ... \cap \eta_{n+1}^\perp \subset \Gamma .$$

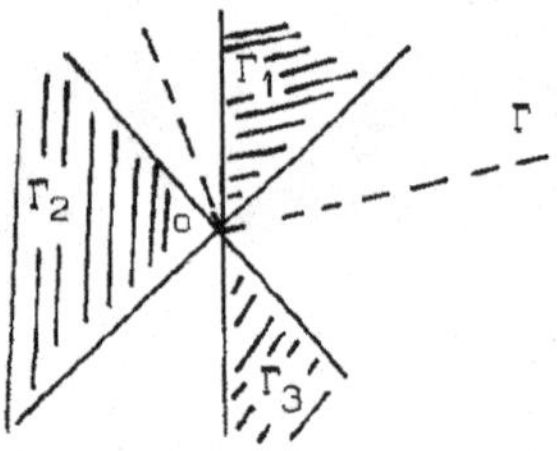

Considérons l'hyperfonction f sur Ω représentée, au moyen de (**) , par les fonctions

$$\Psi_j \in \mathcal{O}(U\cap(\mathbb{R}^n+i\Gamma_j)) \quad (1\leq j\leq n+1)$$

définies ainsi :

$$\begin{cases} \Psi_1 = \varepsilon\, \varphi|_{\Gamma_1} , & \text{avec } \varepsilon = \pm 1 \text{ selon que l'orientation de la base } (\eta_2,..,\eta_{n+1}) \text{ de } \mathbb{R}^n \text{ est positive ou négative} \\ \Psi_j = 0 & \text{si } j=2,..,n+1 \end{cases}$$

Alors, en procédant comme dans l'exemple traité à la fin du §4, on obtient :

THEOREME ET DEFINITION 3

L'hyperfonction $f \in \mathcal{B}(\Omega)$ construite ci-dessus ne dépend pas des vecteurs $\eta_1, \eta_2, .., \eta_{n+1}$ utilisés. On l'appelle valeur au bord de φ selon Γ, et on la note $\varphi(x+i\,\Gamma\, 0)$, ou $b_\Gamma(\varphi)$.

Remarque 2 :

a) Si Γ' est un cône ouvert convexe dans Γ, et si $\varphi \in \mathcal{O}(U\cap(\mathbb{R}^n+i\Gamma))$, il résulte immédiatement de la définition précédente que

$$\varphi(x+i\Gamma' o) = \varphi(x+i\Gamma o)$$

b) De même, si $f \in \mathcal{B}(\Omega)$ est représentée, au moyen de (**) par des fonctions $\Psi_j \in \mathcal{O}(U\cap(\mathbb{R}^n+i\Gamma_j))$, on a

$$f = \sum_{j=1}^{n+1} (-1)^{j-1}\, \Psi_j\, (x+i\Gamma_j\, 0)$$

c) Si $f \in \mathcal{B}(\Omega)$ est représentée, au moyen de (*') par des fonctions $\varphi_\sigma \in \mathcal{O}(U\cap(\mathbb{R}^n+i\Gamma_\sigma))$, on peut vérifier

$$f = \sum_\sigma \varepsilon_\sigma\, \varphi_\sigma\, (x+i\Gamma_\sigma)\ , \text{ où } \varepsilon_\sigma = \sigma_1 \cdots \sigma_n\ ,$$

ce qui généralise la formule donnée dans l'introduction.

Exemple : en utilisant la remarque 2, on peut justifier très simplement le changement de représentation donné à la fin du § 2 pour l'hyperfonction $f = -4\pi\ \delta(x_1, x_2)$. En effet, on a, d'après c)

$$f = b_{\Gamma_{1,1}} \frac{1}{z_1 z_2} - b_{\Gamma_{-1,1}} \frac{1}{z_1 z_2} + b_{\Gamma_{-1,-1}} \frac{1}{z_1 z_2} - b_{\Gamma_{1,-1}} \frac{1}{z_1 z_2}$$

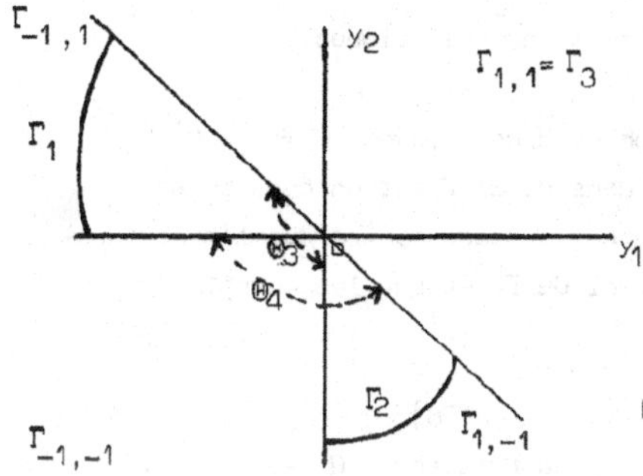

Si Θ_3, Θ_4 sont telles que

$$\begin{cases} \Theta_j \in \mathcal{O}(W_j) \quad (j = 3,4) \\ \dfrac{1}{z_1 z_2} = \Theta_4 - \Theta_3 \quad \text{dans } \Gamma_{-1,-1} \end{cases}$$

on a

$$b_{\Gamma_{-1,-1}} \frac{1}{z_1 z_2} = b_{\Gamma_{-1,-1}} \Theta_4 - b_{\Gamma_{-1,-1}} \Theta_3\ ,$$

et donc, d'après a) :

$$f = b_{\Gamma_1}\left(\frac{-1}{z_1 z_2} - \Theta_3\right) - b_{\Gamma_2}\left(\frac{1}{z_1 z_2} - \Theta_4\right) + b_{\Gamma_3} \frac{1}{z_1 z_2})$$

et b) montre alors que f est représentée par

$$\Psi_1 = \frac{-1}{z_1 z_2} - \Theta_3\ ,\ \Psi_2 = \frac{1}{z_1 z_2} - \Theta_4\ ,\ \Psi_3 = \frac{1}{z_1 z_2}\ .$$

Les parties b) et c) de la remarque 2 montrent que toute hyperfonction est une somme finie de valeurs au bord - Plus généralement, on peut établir le

THEOREME 4

Soient $\Gamma_{1,\ldots,}\Gamma_m$ de cône convexes ouverts de $\mathbb{R}^n$ tels que $\Gamma_{1,\ldots,}^{\perp}\Gamma_m^{\perp}$ recouvrent $\mathbb{R}^n$ (on a posé $\Gamma_j^{\perp} = \{\eta \in \mathbb{R}^n \mid \xi.\eta \geq o \ \forall \xi \in \Gamma_j\}$).

Soient Ω un ouvert de $\mathbb{R}^n$, $f \in \mathcal{B}(\Omega)$, et U un voisinage complexe d'holomorphie de Ω tel que $U \cap \mathbb{R}^n = \Omega$. Alors il existe des fonctions $\varphi_j \in \mathcal{O}(U \cap (\mathbb{R}^n + i\Gamma_j))$ $(j = 1,..,m)$ telles que

$$f = \sum_{j=1}^{m} \varphi_j (x + i \Gamma_j o)$$

Enfin on démontre le théorème suivant, du type "edge of the wedge" :

THEOREME 5

On suppose U connexe. Alors le morphisme "valeur au bord"

$$b_\Gamma : \mathcal{O}(U\cap(\mathbb{R}^n + i\Gamma)) \longrightarrow \mathcal{B}(\Omega) \text{ est injectif}$$

VI - INJECTIONS de $\mathcal{G}$, $\mathcal{D}'$ dans $\mathcal{B}$.

Désignons par $\mathcal{G}$ le faisceau (sur $\mathbb{R}^n$) des fonctions analytiques.
Soient Ω un ouvert de $\mathbb{R}^n$, et $\varphi \in \mathcal{G}(\Omega)$.
Considérons un voisinage complexe d'holomorphie U de Ω (avec $U\cap \mathbb{R}^n = \Omega$) tel que φ se prolonge en fonction holomorphe $\tilde{\varphi}$ dans U. Si Γ est un cône ouvert convexe (de sommet o) arbitraire de $\mathbb{R}^n$, ce qui précède montre que l'hyperfonction $\tilde{\varphi}(x + i\Gamma o)$ sur Ω ne dépend ni de U, ni de Γ, et que les morphismes

$$\mathcal{G}(\Omega) \longrightarrow \mathcal{B}(\Omega)$$
$$\varphi \longmapsto \tilde{\varphi}(x + i\Gamma o)$$

sont injectifs ; on obtient ainsi une injection de faisceaux $\mathcal{G} \longrightarrow \mathcal{B}$.

Soit maintenant $T \in \mathcal{E}'(\mathbb{R}^n)$. La fonction $\varphi \in \mathcal{O}((\mathbb{C} - \mathbb{R})^n)$ définie par $\varphi(z_1,\ldots,z_n) = \frac{1}{(2i\pi)^n} < T(t_1,\ldots t_n), \frac{1}{(t_1-z_1)\ldots(t_n-z_n)} >$ représente, au moyen de (*) une hyperfonction $f_t \in \mathcal{B}(\mathbb{R}^n)$, et on montre que le support de la distribution T est égal au support de l'hyperfonction f_T. Soient maintenant Ω un ouvert de $\mathbb{R}^n$, et $T \in \mathcal{D}'(\Omega)$; décomposons T en une somme localement finie $\sum_j T_j$ de distributions $T_j \in \mathcal{E}'(\mathbb{R}^n)$; la propriété précédente des supports montre que $\sum f_{T_j}$ est aussi localement finie et définit un élément

f_T de $\mathcal{B}(\Omega)$ qui ne dépend pas de la décomposition utilisée pour T, et enfin que les morphismes $\mathfrak{D}'(\Omega) \longrightarrow \mathcal{B}(\Omega)$ sont injectifs

$$T \longmapsto f_T$$

on obtient ainsi une <u>injection</u> de faisceaux $\mathfrak{D}' \longrightarrow \mathcal{B}$, qui prolonge l'injection $\mathcal{C} \longrightarrow \mathcal{B}$. On a donc les sous-faisceaux $\mathcal{C} \subset \mathfrak{D}' \subset \mathcal{B}$.

CHAPITRE III - FAISCEAU $\mathcal{C}$

J. CHAZARAIN

Dans ce chapitre, on expose pour les hyperfonctions à un nombre quelconque de variables, les notions de spectre singulier et de faisceau des singularités qui ont été introduites à la fin du chapitre I dans le cas d'une variable.

Sauf indication contraire, on suppose toujours que M est un ouvert de $\mathbb{R}^n$, un voisinage complexe de M est alors un ouvert X de $\mathbb{C}^n$ tel que M soit fermé dans X. D'autre part, Γ désignera toujours un cône ouvert convexe et $\Gamma^{\perp}$ son cône dual = $\{\eta \mid <\xi,\eta> \geq 0 \text{ tout } \xi \in \Gamma\}$

I - SPECTRE SINGULIER D'UNE HYPERFONCTION.

On a défini au chapitre précédent l'injection i des fonctions analytiques réelles dans les hyperfonctions, de sorte que l'on a une suite exacte de faisceaux sur M

$$0 \longrightarrow \mathcal{A} \xrightarrow{\ i\ } B .$$

Rappelons brièvement la définition de l'hyperfonction $i(\varphi) \in B(M)$ associée à une fonction analytique $\varphi \in \mathcal{A}(M)$. Soit X un voisinage complexe de M où φ se prolonge en une fonction holomorphe et soit Γ un cône convexe ouvert de $\mathbb{R}^n$, on pose

$$f = i(\varphi) = \varphi(x + i\,\Gamma\,0) \quad \text{avec } \varphi \in \mathcal{O}(X \cap \mathbb{R}^n + i\Gamma)$$

et on admis au chapitre II que c'était indépendant au choix de Γ.

Considérons une hyperfonction de la forme

$$f = \sum_{\text{finie}} \varphi_j\,(x + i\,\Gamma_j\,0) \quad \text{avec } \varphi_j \in \mathcal{O}(X \cap \mathbb{R}^n + i\Gamma_j),$$

alors si les φ_j sont holomorphes au voisinage de $x_0 \in M$, f est analytique dans un voisinage de x_0. Ceci est résumé dans la :

PROPOSITION 1.1. - Soit $f \in B(M)$, cette hyperfonction est analytique au voisinage de $x_0 \in M$, si et seulement si elle peut s'écrire

$$f = \sum_{\text{finie}} \varphi_j (x + i\,\Gamma_j\, 0) \quad \text{avec } \varphi_j \in \mathcal{O}(X \cap \mathbb{R}^n + i\,\Gamma_j)$$

avec des φ_j holomorphes dans un voisinage de x_0.

D'autre part, on sait que toute hyperfonction $f \in B(M)$ peut s'écrire sous la forme :

$$f = \sum_{\text{finie}} \varphi_j (x + i\,\Gamma_j\, 0) \quad \text{avec } \varphi_j \in \mathcal{O}(X \cap \mathbb{R}^n + i\,\Gamma_j)$$

pour des cônes Γ_j convenables (cf. chapitre II).

On est alors conduit à poser la :

DEFINITION 1.1. - Soit un covecteur non nul $\eta_0 \in (\mathbb{R}^n)^*$ et soit $x_0 \in M$. On dit qu'une hyperfonction $f \in B(M)$ est micro-analytique au voisinage du point $(x_0, i\eta_0)$ si elle peut s'écrire :

$$f = \sum_{\text{finie}} \varphi_j (x + i\Gamma_j\, 0) \text{ avec } \varphi_j \in \mathcal{O}(X \cap \mathbb{R}^n + i\,\Gamma_j)$$

où les φ_j sont holomorphes au voisinage de x_0 si le cône Γ_j vérifie $\Gamma_j \subset \{\xi \mid \langle \xi, \eta_0 \rangle \geq 0\} = \{\eta_0\}^{\perp}$.

Remarque 1.1. - On vérifie immédiatement que l'expression "f est micro-analytique au voisinage de $(x_0, i\eta_0)$" peut se reformuler en disant que sur un voisinage ouvert de x_0 dans M, on a

$$f = \sum_{\text{finie}} \varphi_j (x + i\,\Gamma_j\, 0)$$

où tous les Γ_j vérifient

$$\Gamma_j \subset \{\xi \mid \langle \xi, \eta_0 \rangle < 0\} .$$

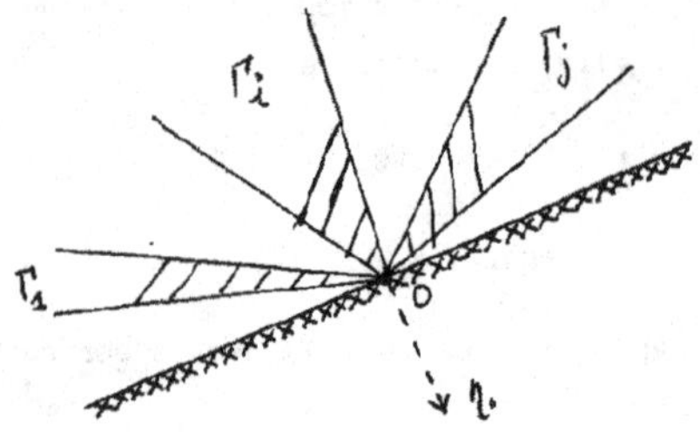

Remarque 1.2. - Il est clair que cette définition coïncide avec celle déjà donnée dans le cas de une variable.

Notons que ce qui importe, ce n'est pas tant le covecteur $\eta_0 \in (\mathbb{R}^n)^* - \{0\}$, que la demi-droite qu'il définit que l'on note η_0^∞ et que l'on considère comme élément de la cosphère $(\mathbb{R}^n)^* - \{0\}/(\text{homothéties} > 0) = S^*$.

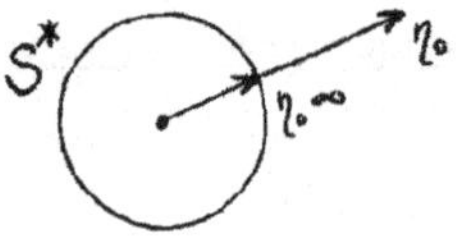

D'autre part, le plongement $M \subset \mathbb{R}^n \subset \mathbb{C}^n = \mathbb{R}^n + i\,\mathbb{R}^n$ permet d'interpréter les covecteurs de la forne $i\eta$ comme des éléments du fibré conormal à M dans $\mathbb{C}^n$ noté $T^*_M\,\mathbb{C}^n$. De façon plus précise, étant donné $x \in M$ on a une suite exacte d'espaces vectoriels

$$0 \longleftarrow T^*_x\,\mathbb{R}^n \xleftarrow{\ \rho\ } T^*_x\,\mathbb{C}^n \longleftarrow \boxed{(T^*_M\,\mathbb{C}^n)_x} \longleftarrow 0 \quad .$$

$$\eta_1 \longleftarrow (\eta_1 + i\eta_2) \qquad \uparrow\ (\text{définition})$$

Ce qui permet d'identifier $T^*_M\,\mathbb{C}^n \simeq M \times i\,(\mathbb{R}^n)^*$ et en passant aux fibrés en cosphères associés, on obtient :

$$S^*_M\,\mathbb{C}^n \simeq M \times i\,S^* = i\,S^*\,M$$

qui est appelé le fibré en cosphères conormales.

On appelle π la projection canonique

$$\begin{array}{ccc} (x, & i\,\eta) \in i\,S^{*}M \\ \downarrow & \downarrow \pi \\ x & M \end{array}$$

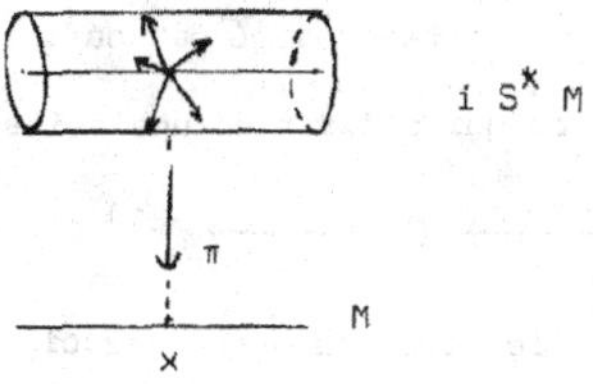

On peut alors poser la :

DEFINITION 1.2. - Soit $f \in B(M)$, l'ensemble des $(x, i\eta\,\infty) \in i\,S^{*}M$ où f n'est pas micro-analytique est un fermé de $i\,S^{*}M$; on l'appelle spectre singulier de f et on le note S.S.f.

Nota : On dit aussi : support singulier ($S.S_C$), support spectral, support essentiel, analytic wave front set.

II - FAISCEAU $\mathcal{C}$.

On commence par définir le faisceau des microfonctions analytiques ; les ouverts de la forme $U \times i\Gamma$ où U est un ouvert de M et Γ désigne la trace sur S^{*} d'un cône ouvert convexe (encore noté Γ) de $(\mathbb{R}^n)^{*}$, forment une base d'ouverts de $i\,S^{*}M$. A l'ouvert $U \times i\Gamma$ on associe

$$\mathcal{A}^{*}(U \times i\Gamma) = \left\{ \begin{array}{l} \text{Le sous espace des hyperfonctions } f \in B(U) \\ \text{telles que S.S.} f \cap (U \times i\Gamma) = \emptyset \end{array} \right\}$$

et soit $\mathcal{A}^{*}$ le faisceau associé sur $i\,S^{*}M$, on l'appelle le faisceau des microfonctions analytiques (ses sections au dessus de $U \times i\Gamma$ sont bien celles ci-dessus).

D'après la définition de l'image inverse d'un faisceau (cf. chapitre I), il est clair que $(\pi^{-1}B)(U \times i\Gamma) = B(U)$, ce qui permet d'interprèter $\mathcal{A}^{*}$ comme un sous-faisceau de $\pi^{-1}B$:

$$0 \longrightarrow \mathcal{A}^{*} \longrightarrow \pi^{-1}B.$$

On définit alors le faisceau $\mathcal{C}$ en complétant cette suite exacte sur $i\,S^{*}M$:

DEFINITION 2.1. - Le faisceau $\mathcal{C}$ est défini sur $i\,S^* M$ comme le quotient de $\pi^{-1} B$ par $\mathcal{A}^*$, ce qui s'écrit en une suite exacte :

$$0 \longrightarrow \mathcal{A}^* \longrightarrow \pi^{-1} B \longrightarrow \mathcal{C} \longrightarrow 0 \text{ (sur } i\,S^* M)$$

C'est à dire que le faisceau $\mathcal{C}$ est associé au préfaisceau défini par les quotients

$$\mathcal{C}'\,(U \times i\Gamma\,) = B(U) \,/\, \mathcal{A}^*\,(U \times i\Gamma\,)$$

où l'on "tue" les hyperfonctions micro-analytiques dans les codirections de Γ, reste donc les singularités.

On a alors le résultat fondamental suivant que l'on admettra

THEOREME 2.1 - Le faisceau $\mathcal{C}$ est flasque.

On projette ensuite la suite exacte précédente sur M par l'application π et l'on démontre que l'on a encore une suite exacte

$$0 \longrightarrow \mathcal{A} \longrightarrow B \xrightarrow{\ sp\ } \pi_* \mathcal{C} \longrightarrow 0,$$

ce qui signifie que $B/\mathcal{A} \simeq \pi_* \mathcal{C}$.

Le morphisme de faisceaux, noté sp, permet de définir l'application

$$f \in B(M) \longrightarrow sp\, f \in (\pi_* \mathcal{C})\,(M) = \mathcal{C}\,(i\,S^* M)$$

et en prenant le support de cette section du faisceau $\mathcal{C}$ on obtient la

PROPOSITION 2.1. - Soit $f \in B(M)$, alors on a

$$\text{supp.}\,(sp\, f) = S.S.f$$

DEMONSTRATION. - Soit $z \in i\,S^* M$, on a les équivalences suivantes :

$z \notin \text{supp.}(spf) \Longleftrightarrow$ il existe un voisinage $\mathcal{V} = U \times i\Gamma$ de z tel que $(spf)/_{\mathcal{V}} = 0$

$\Longleftrightarrow$ il existe un voisinage $\mathcal{V}' = U' \times i\,\Gamma'$ de z tel que
$f/_{U'} \in \mathcal{A}^*(U' \times i\,\Gamma')$

$\Longleftrightarrow$ $z \notin$ S.S.f.

COROLLAIRE 2.1. - Soit $f \in B(M)$, alors on a

$$\text{supp. sing}_a\, f = \pi(\text{S.S.}f)$$

DEMONSTRATION. - On sait que $B/\mathcal{A} \simeq \pi_* \mathcal{C}$, or le support de f dans le faisceau $B/\mathcal{A}$ est précisément le support singulier analytique de f et comme d'autre part, le support dans $\pi_* \mathcal{C}$ est la projection par π de supp. spf, on en déduit le corollaire.

Remarque 2.1. - Bien entendu, SATO donne aussi une définition plus intrinsèque du faisceau $\mathcal{C}$, mais elle nécéssite une grosse machinerie d'algèbre homologique.

Remarque 2.2. - Dans le cadre des fonctions C^∞ et des distributions, HÖRMANDER (Fourier Integral Operater I, Acta Math. 1971) a introduit par anologie un faisceau des singularités à partir du préfaisceau suivant sur $i\,S^* M$, au dessus de l'ouvert $U \times i\Gamma$ on prend

$$\mathcal{D}'(U) \,/\, \{f \in \mathcal{D}'(U) \quad \text{et} \quad WF(f) \cap \Gamma = \emptyset\}.$$

III - FAISCEAU $\mathcal{C}$ ET VALEURS AU BORD.

Pour préciser le lien entre le faisceau $\mathcal{C}$ et l'opération de valeur au bord, on va introduire un nouveau faisceau sur le fibré en sphères normales $i\,S\,M$ (= $M \times i\,S$).

On définit un préfaisceau sur $i\,S\,M$ en posant pour tout ouvert de la forme $U \times i\Gamma \subset M \times i\,S$

$$\tilde{\mathcal{A}}'(U \times i\Gamma) = \lim_{\substack{\longrightarrow \\ X \supset M}} \text{ind}\ \mathcal{O}(X \cap (\mathbb{R}^n + i\Gamma'))$$

où la limite inductive est prise pour les voisinages complexes X de U dans $\mathbb{C}^n$.

Soit $\tilde{\mathcal{Q}}$ le faisceau associé sur $i\,S\,M$, on l'appelle le faisceau des "valeurs au bord idéales de fonctions holomorphes".

On démontre que les sections de $\tilde{\mathcal{Q}}$ au dessus d'un ouvert de la forme $U \times i\Gamma$ sont données par l'expression

$$\tilde{\mathcal{Q}}(U \times i\Gamma) = \lim_{\substack{\longleftarrow \\ \Gamma' \subset\subset \Gamma}} \text{proj} \; \lim_{\substack{\longrightarrow \\ X \supset M}} \text{ind} \; \mathcal{O}(X \cap (\mathbb{R}^n + i\Gamma'))$$

où la limite projective est prise pour les cônes Γ' à base relativement compacte dans Γ. Autrement dit, un élément $f \in \tilde{\mathcal{Q}}(U \times i\Gamma)$ est défini pour tout $\Gamma' \subset\subset \Gamma$ par la donnée d'une fonction holomorphe $f \in \mathcal{O}(X \cap (U \times i\Gamma'))$ où X est un certain voisinage complexe de U qui dépend de f et de Γ', ce que l'on indique par le shéma ci-dessous :

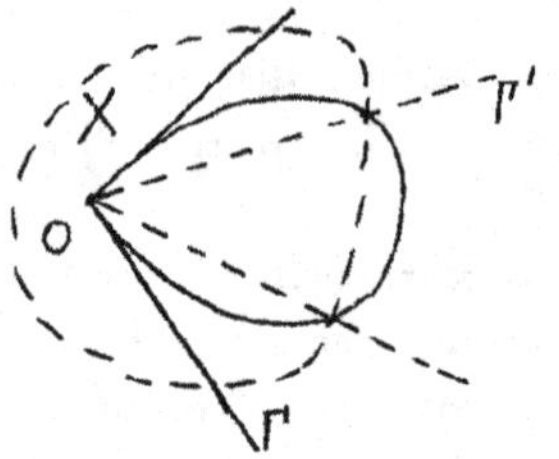

(dessin dans l'espace des parties imaginaires).

L'application "valeur au bord" passe à la limite

$$\begin{array}{ccc} \tilde{\mathcal{Q}}(U \times i\Gamma) & \overset{b}{\dashrightarrow} & B(U) = (\tau^{-1} B)(U \times i\Gamma') \\ \uparrow & \nearrow_{b_{\Gamma'}} & \\ \mathcal{O}(X \cap (U \times i\Gamma')) & & \end{array}$$

et définit un morphisme de faisceaux sur $i\,S\,M$ qui est encore injectif :

$$0 \longrightarrow \tilde{\mathcal{Q}} \xrightarrow{\;b\;} \tau^{-1} B.$$

On a alors le résultat fondamental suivant

<u>THEOREME 3.1.</u> - Soit Γ un côné ouvert convexe et $\Gamma^{\perp}$ son dual.
Soit $f \in B(U)$, on a l'équivalence entre a) et b)

a) Il existe $\varphi \in \tilde{\mathcal{Q}}(U \times i\Gamma)$ telle que $f = b\,\varphi$

b) $S.S.f \subset U \times i(\Gamma^{\perp})$.

Il est utile de disposer d'une forme plus générale de ce théorème relative à des ouverts Z quelconques de i S M.

Auparavant, on définit pour $Z \subset i\,S\,M$ son <u>orthogonal</u> $Z^{\perp}$ dans $i\,S^{*}M$ en posant

$$Z^{\perp} = \{(x,\ i\eta\ \infty) \mid \langle \xi, \eta \rangle \geq 0 \quad \text{tout } \xi \text{ tel que } (x,\ i\xi\ 0) \in Z \}$$

Une partie $Z \subset i\,S\,M$ est dite <u>convexe</u> si chaque fibre $\tau^{-1}(x) \cap Z$ est la trace sur S d'un cône convexe de $\mathbb{R}^n$.

On peut alors énoncer l'important

<u>THEOREME 3.2.</u> - Soit Z un ouvert convexe de i S M et soit $f \in B(\tau Z)$. On a l'équivalence entre a) et b) :

a) Il existe $\varphi \in \tilde{\mathcal{a}}(Z)$ telle que $b\varphi = f$

b) $S.S.\ f \subset Z^{\perp}$.

<u>IV - EXEMPLES.</u>

Pour illustrer ces derniers théorèmes, terminons par quelques exemples.

<u>EXEMPLE 1.</u> -

$$\frac{1}{\langle x,a \rangle + i0}$$

Pour définir d'hyperfonction $f = \dfrac{1}{\langle x,a \rangle + i0} \in B(\mathbb{R}^n)$

(où $\langle x,a \rangle = \sum_{1}^{n} x_j a_j$) on procède de la façon suivante :

la fonction $\varphi(z) = \dfrac{1}{\langle z,a \rangle} = \dfrac{1}{\langle x,a \rangle + i\langle y,a \rangle}$

est holomorphe dans l'ouvert $\tilde{Z} \subset \mathbb{C}^n$ qui est défini par

$$\tilde{Z} = \left\{ z \in \mathbb{R}^n + i\,\mathbb{R}^n \;\middle|\; \begin{array}{l} \langle x,a \rangle \neq 0 \\ \langle x,a \rangle = 0 \text{ et } \langle y,a \rangle > 0 \end{array} \right\}$$

et soit Z la trace sur i S M de cet ouvert conique, alors $\varphi \in \tilde{\mathcal{a}}(Z)$ et on pose $f = b\varphi \in B(\mathbb{R}^n)$.

De plus le théorème 3.2., donne S.S.$f \subset Z^{\perp}$, c'est à dire

$$\text{S.S.}\ \left(\frac{1}{\langle x,a\rangle + i0}\right) \subset \{(x, ia\infty) \mid \langle x,a\rangle = 0\}.$$

Donnons une généralisation utile de cet exemple.

EXEMPLE 2. -

$$\frac{1}{\theta(x) \pm i0}.$$

Soit $\theta : \mathbb{R}^n \longrightarrow \mathbb{R}$, une fonction analytique réelle telle que $d\theta(x) \neq 0$ quand $\theta(x) = 0$ (où $d\theta = (\frac{\partial\theta}{\partial x_1}, \ldots, \frac{\partial\theta}{\partial x_n})$).

Alors on a $f_{\pm} = \dfrac{1}{\theta(x) \pm i0} \in B(\mathbb{R}^n)$ et

$$\text{S.S.}f_{\pm} \subset \{(x, \pm i\, d\theta(x)\infty) \mid \theta(x) = 0\}.$$

En effet, on sait que la fonction θ admet un prolongement holomorphe dans un voisinage complexe X de $\mathbb{R}^n$. On pose

$$\theta(z) = \text{Re}\,\theta(z) + i\,\mathfrak{Im}\,\theta(z)$$

et on applique la formule de Taylor entre $x + iy$ et x :

$$\text{Re}\ \theta(x + iy) = \theta(x) + i \langle y, d\theta(x)\rangle + |y|\,\varepsilon(x)$$

$$\mathfrak{Im}\ \theta(x + iy) = |y|\,\varepsilon(x)$$

d'où

$$\theta(z) = \theta(x) + i\langle y, d\theta(x)\rangle + |y|\,\varepsilon(x).$$

Par conséquent la fonction $\varphi(z) = \dfrac{1}{\theta(z)}$ est holomorphe dans l'ouvert

$$\tilde{Z}_{\pm} = \left\{ z = x + iy \in \mathbb{C}^n \;\middle|\; \begin{array}{l} \theta(x) \neq 0 \\ \text{ou} \\ \theta(x) = 0 \text{ et } \langle y, d\ (x)\rangle \gtrless 0 \end{array} \right\} \cap X'$$

où X' est un "petit" voisinage complexe de $\mathbb{R}^n$.

Alors l'assertion sur le support singulier de $f = b_{\pm}\varphi$ découle du théorème 3.2.

EXERCICE : Appliquer ceci à l'hyperfonction $\dfrac{1}{p^2 - m^2 + i0}$,

EXEMPLE 3.

$$\delta(\varphi).$$

Avec les mêmes conditions sur $\varphi\ : \mathbb{R}^n \longrightarrow \mathbb{R}$ on définit

$$\delta(\varphi) = \frac{-1}{2\pi i}\left(\frac{1}{\varphi + i0} - \frac{1}{\varphi - i0}\right) \in \mathcal{B}(\mathbb{R}^n)$$

ce qui est naturel d'après la représentation de

$$\delta(t) = \frac{-1}{2\pi i}\left(\frac{1}{t + i0} - \frac{1}{t - i0}\right) \in \mathcal{B}(\mathbb{R}).$$

Et d'après l'exemple 2, on a

$$\text{S.S. } \delta(\varphi) \subset \{ (x, \pm i d\varphi(x)\infty) \mid \varphi(x) = 0 \}.$$

RESUME DU CHAPITRE

On a les suites exactes :

$$\begin{array}{ccc} i\,S\,M & & i\,S^*M \\ & \tau \searrow \quad \swarrow \pi & \\ & M & \end{array}$$

$$0 \longrightarrow \mathcal{A} \longrightarrow \mathcal{B} \longrightarrow \pi_*\mathcal{C} \longrightarrow 0 \quad \text{(sur } M)$$

$$0 \longrightarrow \mathcal{A}^* \longrightarrow \pi^{-1}\mathcal{B} \xrightarrow{sp} \mathcal{C} \longrightarrow 0 \quad \text{(sur } i\,S^*M)$$

$$0 \longrightarrow \tilde{\mathcal{A}} \xrightarrow{b} \tau^{-1}\mathcal{B} \longrightarrow \boxed{} \longrightarrow 0 \quad \text{(sur } i\,S\,M).$$

Soit Z un ouvert convexe de $i\,S\,M$ et $f \in \mathcal{B}(\tau Z)$, alors

$$\left\{\begin{array}{l} f = b\varphi \\ \varphi \in \tilde{\mathcal{A}}(Z) \end{array}\right. \Longleftrightarrow \text{S.S.} f \subset Z^{\perp}$$

CHAPITRE IV - APPLICATIONS

J. CHAZARAIN

Dans ce chapitre, on expose la formulation hyperfonction du théorème du "Edge of the wedge", puis on décrit les opérations usuelles sur les hyperfonctions (multiplication, restriction,...) que l'on peut définir grâce à des hypothèses sur le spectre singulier.

I - EDGE OF THE WEDGE.

Commençons par une formulation avec deux cônes :

THEOREME 1.1. : Soient Γ_1, Γ_2 des cônes ouverts convexes de $\mathbb{R}^n - \{0\}$. Soient $\varphi_j \in \tilde{\mathcal{a}}(\mathbb{R}^n + i\Gamma_j)$ $j = 1,2$ telles que $b\varphi_1 = b\varphi_2$, alors il existe $\varphi \in \tilde{\mathcal{a}}(\mathbb{R}^n + i\,\widehat{\Gamma_1 \cup \Gamma_2})$ telle que φ prolonge φ_1 et φ_2. En particulier si $\Gamma_1 = -\Gamma_2$, alors φ est holomorphe dans un voisinage de $\mathbb{R}^n$.

($\widehat{A}$ désigne l'enveloppe convexe de la partie A).

DEMONSTRATION : Posons $f = b\varphi_1 = b\varphi_2 \in \mathcal{B}(\mathbb{R}^n)$, le théorème 3.1 (chap.III) montre que :

S.S. $f \subset (\mathbb{R}^n + i\Gamma_1^{\circ}) \cap (\mathbb{R}^n + i\Gamma_2^{\circ}) = \mathbb{R}^n + i\,(\widehat{\Gamma_1 \cup \Gamma_2})^{\perp}$ donc on a $f = b\varphi$ avec $\varphi \in \tilde{\mathcal{a}}(\mathbb{R}^n + i\,\widehat{\Gamma_1 \cup \Gamma_2})$.

Et d'après l'injectivité de b, on en déduit que φ prolonge φ_1 et φ_2.

Pour énoncer la généralisation avec un nombre fini de cônes, on introduit la notion de partie <u>propre</u> de $i\,S\,M$. On dit que $Z \subset i\,S\,M$ est propre si $Z \cap Z^{a} = \emptyset$, où Z^{a} désigne l'image de Z par l'application antipodale sur $i\,S\,M$:

$$a : (x, i\xi) \longrightarrow (x, -i\xi).$$

On a le

<u>THEOREME 1.2</u>. (Edge of the wedge). Soient des cônes ouverts convexes propres de $i\,S\,M$: $Z_1, \dots, Z_N$ et soient des $\varphi_j \in \tilde{\mathcal{Q}}(Z_j)$ telles que $\sum_{j=1}^{N} b\,\varphi_j = 0$.

Alors il existe des $\varphi_{j,k} \in \tilde{\mathcal{Q}}(Z_j \cup Z_k)$ telles que

$$\varphi_j = \sum_{k=1}^{N} \varphi_{j,k} \qquad \text{pour tout } j = 1,\dots,N$$

et

$$\varphi_{j,k} = -\varphi_{k,j} \qquad \text{tout } j,k$$

Avant d'indiquer la démonstration de ce théorème, indiquons aussi la

<u>PROPOSITION 1.1</u>. - Soit $f \in B(M)$ et supposons que $S.S.\, f \subset \bigcup_{j=1}^{N} Z_j$ où les Z_j sont des ouverts convexes propres de $i\,S\,M$. Alors il existe des $\varphi_j \in \tilde{\mathcal{Q}}(Z_j)$ telles que $f = \sum_{1}^{N} b\,\varphi_j$

Ceci posé, le théorème 1.2 et la proposition 1.1. sont (ainsi que me l'a indiqué M. Kashiwara) des corollaires immédiats du résultat suivant de théorie des faisceaux.

<u>THEOREME 1.3.</u> - Soit $\mathcal{F}$ un faisceau flasque sur un espace X, et soient des fermés F_j $j = 1,\dots,N$ de X. Alors on a la suite exacte de groupes

$$0 \longleftarrow \Gamma_{\cup F_j}(X,\mathcal{F}) \xleftarrow{\ \alpha\ } \prod_j \Gamma_{F_j}(X,\mathcal{F}) \xleftarrow{\ \beta\ } \prod_{j,k}{}' \Gamma_{F_j \cap F_k}(X,\mathcal{F})$$

où $\alpha : (\varphi_j)_j \longrightarrow \sum_j^N \varphi_j$

$\beta : (\varphi_{j,h})_{j,h} \longrightarrow (\sum_{k=1}^N \varphi_{j,h})_j$

et Π' désigne l'espace des $(\varphi_{j,h})_{j,h}$ avec $\varphi_{j,h} + \varphi_{h,j} = 0$

Démonstration de ce théorème :

1 - Démonstration de la surjectivité de α.

On procède par récurrence sur N. Commençons par N = 2, soit donc $\varphi \in \Gamma_{F_1 \cup F_2}(X,\mathcal{F})$ et construisons $\varphi_j \in \Gamma_{F_j}(X,\mathcal{F})$ telles que $\varphi = \varphi_1 + \varphi_2$.

On pose

$$\varphi_1' = \begin{cases} \varphi & \text{sur } X \setminus F_2 \\ 0 & \text{sur } X \setminus F_1 \end{cases}$$

Il est clair que c'est compatible sur $X \setminus (F_1 \cap F_2)$ et comme $\mathcal{F}$ est flasque, on peut prolonger φ_1' en une section $\varphi_1 \in \Gamma_{F_1}(X,\mathcal{F})$. On pose ensuite $\varphi_2 = \varphi - \varphi_1$ et on vérifie immédiatement que $\varphi_2 \in \Gamma_{F_2}(X,\mathcal{F})$.

Dans le cas N quelconque, on se ramène immédiatement au cas N = 2 en écrivant $F_1 \cup \ldots \cup F_N = (F_1 \cup \ldots \cup F_{N-1}) \cup F_N$ et on termine grâce à l'hypothèse de récurrence.

2 - Démonstration de l'exactitude de la suite en $\prod_j \Gamma_{F_j}(X,\mathcal{F})$.

Désignons par (β_N) la propriété de β d'être exacte quand on a N fermés. On démontre que l'on a (β_N) pour tout N en procédant par récurrence sur N.

Pour N = 2, c'est immédiat car l'hypothèse $\varphi_1 + \varphi_2 = 0$ implique $\varphi_1 = -\varphi_2 \in \Gamma_{Z_1 \cap Z_2}(X,\mathcal{F})$ ce qui permet de poser $\varphi_{12} = \varphi_1$ et $\varphi_{21} = \varphi_2$.

Montrons que $(\beta_{N-1}) \Longrightarrow (\beta_N)$.

Soit donc $(\varphi_j)_{j=1,\dots,N}$ telles que $\alpha\,((\varphi_j)) = 0$

On écrit, $\varphi_1 + (\varphi_2 + \dots + \varphi_N) = 0$

d'où $\operatorname{supp.}\varphi_1 \subset \bigcup_{k=2}^{N} (F_1 \cap F_k)$

et la surjectivité de α permet d'écrire

$$\varphi_1 = \sum_{k=2}^{N} \varphi_{1,k} \text{ avec } \varphi_{1,k} \in \Gamma_{Z_1 \cap Z_k}(X,\mathcal{F})$$

et on pose $\varphi_{k,1} = -\varphi_{1,k}$.

Introduisons $\psi_k = \varphi_k + \varphi_{1,k}$ $k = 2,\dots,N$, de sorte que l'on a $\sum_{2}^{N} \psi_k = 0$,
alors l'hypothèse de récurrence permet d'écrire

$$\psi_k = \sum_{j=2}^{N} \varphi_{k,j} \text{ avec } \varphi_{k,j} \in \Gamma_{Z_k \cap Z_j}(X,\mathcal{F}) \text{ et } \varphi_{h,j} + \varphi_{j,h} = 0 \quad j,h \geq 2.$$

On vérifie alors finalement que les $(\varphi_{j,h})_{j,h\,=\,1,\dots,N}$ ainsi construits sont bien antisymétriques en (j,h) et que

$$\varphi_j = \sum_{k=1}^{N} \varphi_{j,k}.$$

La démonstration de la proposition 1.1. découle alors de l'exactitude de α appliquée au faisceau $\mathcal{C}$ et la démonstration du théorème 1.2. découle, compte tenu du théorème 3.2. Chap. III, de l'exactitude de β appliquée avec $\mathcal{F} = \mathcal{C}$.

II - OPERATIONS SUR LES HYPERFONCTIONS.

Il est possible, moyennant des hypothèses sur le spectre singulier, d'étendre aux hyperfonctions certaines opérations usuelles sur les fonctions (multiplication, restriction, substitution, intégration).

a) - Multiplication :

THEOREME 2.2. - Soient des hyperfonctions $f,\ g \in B(M)$ telles que $S.S.f \cap (S.S.g)^a = \emptyset$, alors on peut définir de façon naturelle le produit $f \cdot g \in B(M)$ et on a

$$S.S.(f.g) \subset S.S.f \cup S.S.g \cup (S.S.f + S.S.g).$$

DEMONSTRATION : On recouvre S.S.f (resp S.S.g) par des fermés $Z_j^{\perp}$ (resp. $Z_k'^{\perp}$) où les Z_j, Z'_k sont des ouverts propres convexes de $i\ S\ M$ et tels que $Z_j^{\perp} \cap (Z_k'^{\perp})^a = \emptyset$, ce qui est possible grâce à l'hypothèse sur les spectres singuliers. La proposition 1.1. permet d'écrire les décompositions

$$f = \Sigma\ b\varphi_j \qquad \varphi_j \in \tilde{\mathcal{Q}}(Z_j)$$

$$g = \Sigma\ b\varphi'_k \qquad \varphi'_k \in \tilde{\mathcal{Q}}(Z'_k)$$

et on pose par définition

$$f.g = \sum_{j,k} b\ (\varphi_j . \varphi'_k).$$

Cela a bien un sens car $\varphi_j . \varphi'_k \in \tilde{\mathcal{Q}}(Z_j \cap Z'_k)$ et $Z_j \cap Z'_k \neq \emptyset$ puisque $Z_j^{\perp} \cap (Z_k'^{\perp})^a = \emptyset$.

De plus, cette définition montre que

$$S.S.(f.g) \subset \bigcup_{j,k} \widehat{(Z_j^{\perp} \cup Z_k'^{\perp})}$$

$$\subset \bigcup Z_j^{\perp} \cup Z_k'^{\perp} \cup (Z_j^{\perp} + Z_k'^{\perp}),$$

d'où l'inclusion annoncée pour S.S.(f.g) en raffinant les recouvrements.

Bien entendu, il faut s'assurer que cette définition est indépendante du choix des φ_j, φ'_k. Soit ψ_j un autre choix de φ_j mais relatif au même recouvrement Z_j.

On a donc $\sum_j b(\varphi_j - \psi_j) = f - f = 0$

et le théorème 1.2 permet alors d'écrire

$$\varphi_j - \psi_j = \sum_k \varphi_{j,k} \quad \text{avec } \varphi_{j,k} \in \tilde{\mathcal{Q}}(\widehat{Z_j \cup Z_k})$$

d'où $\quad \sum_{j,k} b(\varphi_j \varphi'_k) - \sum_{j,k} b(\psi_j \varphi'_k)$

$$= \sum_{j,h,k} b(\varphi_{j,h} \cdot \varphi'_k)$$

$$= 0, \text{ car } \varphi_{j,h} + \varphi_{h,j} = 0.$$

L'indépendance par rapport aux recouvrements se démontre en passant à des recouvrements plus fins.

EXEMPLE.

On peut définir $\left(\frac{1}{\Theta(x)+i0}\right)^p = b\left(\frac{1}{\Theta(z)^p}\right)$ avec les notations de l'exemple 2 (chap. III), et on a

$$S.S. \left(\frac{1}{\Theta(x)+i0}\right)^p \subset \{(x, id\Theta(x)\infty)\mid \Theta(x) = 0\} .$$

En revanche, $\big(\delta\ (\Theta)\big)^p$ n'est pas défini pour $p \geq 2$ car l'hypothèse du théorème 2.1. n'est pas satisfaite.

b) - Restriction d'une hyperfonction à une sous-variété.

Soit $N \subset \mathbb{R}^n$, une sous-variété analytique réelle de codimension p, on peut l'écrire localement sous la forme

$$N = \{ x \mid \Theta_1(x) = \ldots = \Theta_p(x) = 0 \}$$

avec des fonctions analytiques réelles Θ_j telles que les différentielles $d\Theta_1(x),\ldots,d\Theta_p(x)$ soient indépendantes quand $x \in N$. Le sous espace de $T^*_x \mathbb{R}^n$ engendré par ces formes, s'appelle l'espace conormal à N en x et est noté $(T^*_N \mathbb{R}^n)_x$. On rappelle la suite exacte :

$$\begin{array}{ccccccccc} 0 & \longleftarrow & T^*_x N & \overset{\rho}{\longleftarrow} & T^*_x \mathbb{R}^n & \longleftarrow & (T^*_N \mathbb{R}^n)_x & \longleftarrow & 0 \quad (x \in N) \\ & & \cup\!\!\!\!| & & \cup\!\!\!\!| & & & & \\ \rho(\eta) & = & \eta\big|_{T_x N} & \longleftarrow & \eta & & & & \end{array}$$

et soient S^*N, $S^*\mathbb{R}^n$, $S^*_N\mathbb{R}^n$ les fibrés en cosphères associés. On peut alors énoncer le

THEOREME 2.2. - Soit N une sous-variété analytique réelle de $\mathbb{R}^n$ définie par des équations $\theta_j(x) = 0$, $j = 1,\dots,P$. Soit $f \in \mathcal{B}(\mathbb{R}^n)$ telle que $S.S.f \cap i\, S^*_N\mathbb{R}^n = \emptyset$, alors on peut définir de façon naturelle la restriction à N de f soit

$f|_N \in \mathcal{B}(N)$ et on a

$$S.S.(f|_N) \subset \{(x, i\rho(\eta)\infty) \mid (x, i\eta\infty) \in S.S.f \text{ et } x \in N\}.$$

Remarque 2.1. - Bien entendu, on peut donner un énoncé intrinsèque de ce théorème mais cela nécessite quelques digressions sur les fibrés en cosphères normales.

Démonstration du théorème. -

C'est un résultat local, il suffit donc de le démontrer quand N est un sous-espace vectoriel de $\mathbb{R}^n$, dans ce cas on a

$$T^*_N\mathbb{R}^n = \{(x,\eta) \mid x \in N \text{ et } \eta \in N^\perp\}.$$

Soit $f \in \mathcal{B}(\mathbb{R}^n)$ telle que $\{(x, i\eta\infty) \in S.S.f \text{ et } x \in N\} \Rightarrow \eta \notin N^\perp$. On recouvre S.S.f par des fermés $Z_j^\perp = (U_j \times i\Gamma_j^\perp)$ où les Γ_j sont des cônes ouverts convexes propres de $\mathbb{R}^n$ tels que $\Gamma_j \cap N^\perp = \emptyset$.

Soit $f = \Sigma\, b\, \varphi_j$ $\varphi_j \in \tilde{\mathcal{Q}}(\mathbb{R}^n + i\Gamma_j)$, une décomposition associée de f. Commençons par démontrer le

LEMME 2.1. - Soit $\varphi \in \tilde{\mathcal{Q}}(\mathbb{R}^n + i\Gamma)$ avec $\Gamma^\perp \cap N^\perp = \emptyset$.

Alors on peut définir de façon naturelle

$$\varphi|_{iSN} \in \tilde{\mathcal{Q}}(N + i(\Gamma \cap N)).$$

DEMONSTRATION. - Pour tout $\Gamma' \subset\subset \Gamma$, il existe, d'après la définition de $\tilde{\mathcal{Q}}(\mathbb{R}^n + i\Gamma')$, un voisinage complexe X de $\mathbb{R}^n$ et $\varphi(z)$ appartenant à $\mathcal{O}(X \cap (\mathbb{R}^n + i\Gamma'))$ qui represente φ. D'autre part, on vérifie que si Γ' est assez voisin de Γ, l'intersection $\Gamma' \cap N$ est un cône non réduit à {0} car $\Gamma^\perp \cap N^\perp = \emptyset$.

Par conséquent, la restriction $\varphi|_{(N + i(N\cap\Gamma'))\cap X}$ définie un élément de $\tilde{\mathcal{A}}(N + i(\Gamma\cap N))$ que l'on note $\varphi|_{iSN}$.

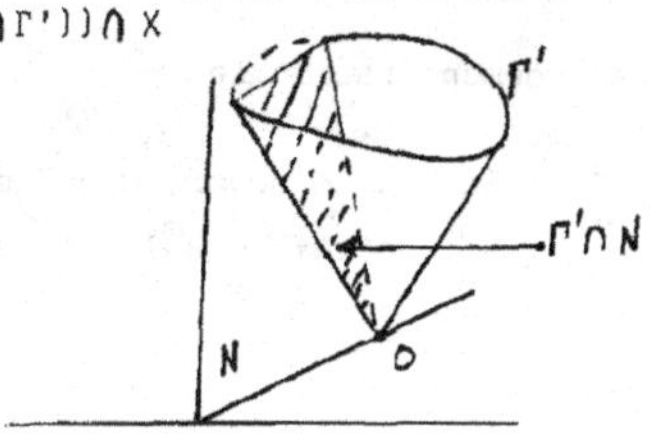

Revenons à f, on pose par définition

$$f|_N = \sum_j b(\varphi_j|_{iSN}) \in B(N).$$

On a donc

$$S.S.\ (f|_N) \subset \bigcup_j (N + i(\Gamma_j\cap N)^{\perp'})$$

où $\perp'$ signifie l'orthogonal dans le sous-espace N, cet orthogonal est égal à $\{p(\eta) \mid \eta\in\Gamma_j^{\perp}\}$, et il vient

$$S.S.\ (f|_N) \subset \{(x, ip(\eta)\infty) \mid (x,i\eta\infty)\in \bigcup_j (\mathbb{R}^n + i\Gamma_j^{\perp}) \text{ et } x\in N.\}$$

d'où le théorème. La démonstration de l'indépendance de cette définition relativement aux Z_j et φ_j se fait comme dans le théorème de multiplication.

Donnons une application de ce théorème à une situation fréquente.
Soit $f \in B(\mathbb{R}^n)$ telle que S.S.f ne contient aucun point de la forme $(x',0;0,\eta_n)$ où $x = (x', x_n)$ et $\eta = (\eta',\eta_n)$, alors on peut définir $f|_{(x_n = 0)}$ et on a

$$S.S.\ (f|_{x_n=0}) \subset \{ (x',i\eta'\infty) \mid \exists\ (x',0;i(\eta',\eta_n)\infty) \in S.S.f \}.$$

Par exemple, on sait que l'hypothèse sur S.S.f est vérifiée quand f vérifie une équation aux dérivées partielles

$$P(x,D)\ f = 0$$

telle que l'hyperplan $x_n = 0$ ne soit pas caractéristique pour l'opérateur P, on peut donc donner un sens aux traces sur $x_n = 0$ de f et de ses dérivées.

c) Changement de variables (ou image réciproque) pour les hyperfonctions.

Enonçons sans démonstration le

THEOREME 2.3. - Soit N un ouvert de $\mathbb{R}^n$ et $\mathcal{O}$ une application analytique réelle $N \xrightarrow{\mathcal{O}} \mathbb{R}^m$. Soit $f \in B(\mathbb{R}^m)$, alors on peut définir de façon naturelle la composée

$$\mathcal{O}^* f \in B(N)$$

sous l'hypothèse : $(\mathcal{O}(x), i\eta\infty) \in S.S.f \Longrightarrow \sum_{j=1}^{m} \eta_j \; d\mathcal{O}_j(x) \neq 0.$

Et on a alors

$$S.S.(\mathcal{O}^* f) \subset \{(x, i(\Sigma\, \eta_j \; d\mathcal{O}_j(x))\infty) \mid (\mathcal{O}(x), i\eta\infty) \in S.S.f\} .$$

EXEMPLE. - Avec $\mathcal{O} : \mathbb{R}^n \longrightarrow \mathbb{R}$ comme à l'exemple 2 (chap. III), on montre que $\mathcal{O}^* \delta_o = \delta(\mathcal{O})$.

Remarque 2.2. - Bien entendu, on peut donner une formulation intrinsèque de ce théorème relatif au cas où $\mathcal{O}$ est une application entre deux variétés analytiques réelles.

d) Intégration (ou image directe) pour les hyperfonctions.

Enonçons sans démonstration le

THEOREME 2.4. - Soient N, M des variétés analytiques réelles et $\mathcal{O}$ une submersion analytique $N \xrightarrow{\mathcal{O}} M$ (c'est-à-dire que l'application tangente $\mathcal{O}'(x)$ est surjective pour tout x). On suppose également que la restriction de $\mathcal{O}$ au support de f est propre (supp. f) $\xrightarrow{\mathcal{O}}$ M. Soit $f \in B(N) \boxtimes \omega_N$ une hyperfonction à valeurs dans les formes de degré maximum sur N, alors on peut définir de façon naturelle son intégrale sur les fibres de $\mathcal{O}$:

$$\text{"} (\mathcal{O}_* f)(x) = \int_{\mathcal{O}^{-1}(x)} f \text{"} \in B(M) \boxtimes \omega_M .$$

Et on a :

$$S.S.\ (\mathcal{O}_* f) \subset \{ (\mathcal{O}(x), i\eta\infty) \mid (x, i\,{}^t\mathcal{O}'(x)\eta\infty) \in S.S.f \}$$

où ${}^t\sigma'(x)\,\eta$ signifie dans une carte de M : $\sum_1^m \eta_j \, d\sigma_j(x)$.

<u>Remarque 2.3.</u> - On peut supprimer l'hypothèse que σ est une submersion, mais il faut alors rajouter à S.S. $(\sigma_* f)$ l'ensemble

$$\{(\sigma(x), i\eta\infty) \mid {}^t\sigma'(x)\eta = 0 \text{ et } x \in \text{supp. } f\}.$$

$$" \int_{\sigma^{-1}(x)} f \; "$$

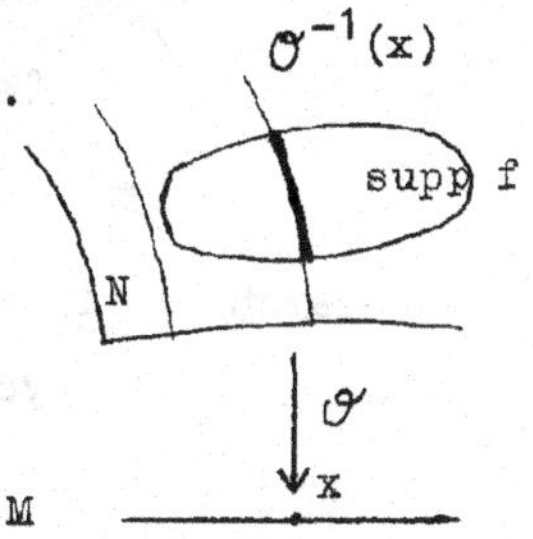

Donnons un cas particulier important de ce théorème, c'est le cas où σ est une projection

$$\begin{array}{ccc} N = T \times M & \ni & (t,x) \\ \downarrow \sigma & & \downarrow \\ M & \ni & x \end{array} .$$

alors $f \in B(N) \otimes \omega_N$ s'écrit formellement $f(t,x)\,dt\,dx$ et on a

$$(\sigma_* f)(x)\,dx = \left(\int_T f(t,x)\,dt\right) dx$$

avec

$$\text{S.S. } (\sigma_* f) \subset \{(x, i\eta\infty) \mid \exists\ (t,x;\, i(0,\eta)\infty) \in \text{S.S.} f\} .$$

Terminons par un

<u>EXEMPLE</u>. - On a la représentation suivante de $\delta_0 \in B(R^n)$ dite "représentation en ondes planes".

$$\delta_0 = \frac{(n-1)\,!}{(2\pi i)^n} \int_{|\xi| = 1} \frac{\omega(\xi)}{(\langle x,\xi\rangle + i0)^n}$$

avec $\omega(\xi) = \sum_{j=1}^n (-1)^j \xi_j \, d\xi_1 \wedge \ldots \wedge \widehat{d\xi_j} \wedge \ldots \wedge d\xi_n$

et où l'intégrale signifie l'intégrale de l'hyperfonction

$$\frac{\omega(\xi)}{(\langle x,\xi\rangle + i0)^n} \in B(R^n \times S^*) \text{ sur la sphère } |\xi| = 1.$$

Cette représentation de δ joue un rôle important dans la définition des opérateurs pseudo-différentiels.

PSEUDO-DIFFERENTIAL OPERATORS ACTING ON THE SHEAF OF MICROFUNCTIONS

Takahiro KAWAI

Research Institute for Mathematical Sciences,
Kyoto University
and
Département de Mathématiques, Nice

This report is intended to give an intuitive explanation of the theory of pseudo-differential operators in hyperfunction theory. The reader is referred to Sato-Kawai-Kashiwara [1] for further details of the theory and its applications. Hereafter Sato-Kawai-Kashiwara [1] will be quoted as S-K-K for short. In this exposition the present speaker wants to lay great emphasis on the following two points.

I. The pseudo-differential operator which we want to manipulate is of infinite order.

II. The pseudo-differential operator acts on the sheaf of microfunctions as a sheaf homomorphism.

The present speaker hopes that tho importance of these properties will be recognized clearly by this exposition.

To begin with, we recall the definition of linear differential operators in hyperfunction theory.

Intuitively speaking, a linear differential operator (with real analytic coefficients) on a real analytic manifold M is a special kind of integral operators whose kernel functions have their support in the diagonal set $\Delta_M=\{(x,y)\in M\times M;\ x=y\}$. (See S-K-K Chapter I §2.1 for the precise definition.) Here we use hyperfunctions as kernel functions. This fact corresponds to the fact that there apper linear differential operators of infinite order. One of the typical examples of such operators is given by

$$\cosh\sqrt{\frac{d}{dx}} = \sum_{n=0}^{\infty} \frac{1}{(2n)!} \left(\frac{d}{dx}\right)^n .$$

One of the essential points in our argument is the following: in order that an infinite sum of differential operators of finite order should make sense as a differential operator of infinite order thus defined, very servere condition should be posed. For example,

$$\exp\left(\frac{d}{dx}\right) = \sum_{n=0}^{\infty} \frac{1}{n!}\left(\frac{d}{dx}\right)^n$$

never gives rise to a differential operator of infinite order. In fact the kernel function corresponding to $\exp(\frac{d}{dx})$ is $\delta(x-y-1)dy$, whose support is not contained in $\Delta_{\mathbb{R}}=\{x=y\}$. The precise condition will be clarified later (see condition (3)) and here we only mention one example which shows the advantage of employing pseudo-differential operators of infinite order.

We have the following relation:

$$\begin{pmatrix} \cosh(-x_1\sqrt{\frac{\partial}{\partial x_2}}) & \frac{1}{\sqrt{\frac{\partial}{\partial x_2}}}\sinh(-x_1\sqrt{\frac{\partial}{\partial x_2}}) \\ \sqrt{\frac{\partial}{\partial x_2}}\sinh(-x_1\sqrt{\frac{\partial}{\partial x_2}}) & \cosh(-x_1\sqrt{\frac{\partial}{\partial x_2}}) \end{pmatrix} \times$$

$$\times \begin{pmatrix} \frac{\partial}{\partial x_1} & -1 \\ -\frac{\partial}{\partial x_2} & \frac{\partial}{\partial x_1} \end{pmatrix} \begin{pmatrix} \cosh(x_1\sqrt{\frac{\partial}{\partial x_2}}) & \frac{1}{\sqrt{\frac{\partial}{\partial x_2}}}\sinh(x_1\sqrt{\frac{\partial}{\partial x_2}}) \\ \sqrt{\frac{\partial}{\partial x_2}}\sinh(x_1\sqrt{\frac{\partial}{\partial x_2}}) & \cosh(x_1\sqrt{\frac{\partial}{\partial x_2}}) \end{pmatrix}$$

$$= \begin{pmatrix} \frac{\partial}{\partial x_1} & 0 \\ 0 & \frac{\partial}{\partial x_1} \end{pmatrix}.$$

Here $\frac{1}{\sqrt{\frac{\partial}{\partial x_2}}}\sinh(x_1\sqrt{\frac{\partial}{\partial x_2}}) = \sum_{n=0}^{\infty} \frac{x_1^{2n+1}}{(2n+1)!}(\frac{\partial}{\partial x_2})^n$ and so on.

Clearly

$$\begin{pmatrix} \cosh(-x_1\sqrt{\frac{\partial}{\partial x_2}}) & \frac{1}{\sqrt{\frac{\partial}{\partial x_2}}}\sinh(-x_1\sqrt{\frac{\partial}{\partial x_2}}) \\ \sqrt{\frac{\partial}{\partial x_2}}\sinh(-x_1\sqrt{\frac{\partial}{\partial x_2}}) & \cosh(-x_1\sqrt{\frac{\partial}{\partial x_2}}) \end{pmatrix} \times$$

$$\times \begin{pmatrix} \cosh(x_1\sqrt{\frac{\partial}{\partial x_2}}) & \frac{1}{\sqrt{\frac{\partial}{\partial x_2}}}\sinh(x_1\sqrt{\frac{\partial}{\partial x_2}}) \\ \sqrt{\frac{\partial}{\partial x_2}}\sinh(x_1\sqrt{\frac{\partial}{\partial x_2}}) & \cosh(x_1\sqrt{\frac{\partial}{\partial x_2}}) \end{pmatrix}$$

$$= \begin{pmatrix} \cosh(x_1\sqrt{\frac{\partial}{\partial x_2}}) & \frac{1}{\sqrt{\frac{\partial}{\partial x_2}}}\sinh(x_1\sqrt{\frac{\partial}{\partial x_2}}) \\ \sqrt{\frac{\partial}{\partial x_2}}\sinh(x_1\sqrt{\frac{\partial}{\partial x_2}}) & \cosh(x_1\sqrt{\frac{\partial}{\partial x_2}}) \end{pmatrix} \times$$

$$\times \begin{pmatrix} \cosh(-x_1\sqrt{\frac{\partial}{\partial x_2}}) & \frac{1}{\sqrt{\frac{\partial}{\partial x_2}}}\sinh(-x_1\sqrt{\frac{\partial}{\partial x_2}}) \\ \sqrt{\frac{\partial}{\partial x_2}}\sinh(-x_1\sqrt{\frac{\partial}{\partial x_2}}) & \cosh(-x_1\sqrt{\frac{\partial}{\partial x_2}}) \end{pmatrix}$$

$$= \begin{pmatrix} 1 & 0 \\ 0 & 1 \end{pmatrix}$$ holds and this

relation implies that

$$\begin{pmatrix} \frac{\partial}{\partial x_1} & -1 \\ -\frac{\partial}{\partial x_2} & \frac{\partial}{\partial x_1} \end{pmatrix} \text{ and } \begin{pmatrix} \frac{\partial}{\partial x_1} & 0 \\ 0 & \frac{\partial}{\partial x_1} \end{pmatrix}$$

are transformed each other by an inner automorphism. Especially this fact tells us that the differential equation $(\frac{\partial^2}{\partial x_1^2} - \frac{\partial}{\partial x_2})u = 0$ is equivalent to the equation $\frac{\partial^2}{\partial x_1^2} u = 0$. Such a transparent result can be obtained only after the introduction of differential operators of infinite order. The enlarged version of this fact plays its essential role in discussing the theory of general system of pseudo-differential equations. (See S-K-K Chapter II §5.3 and Chapter III §2.) Here we would like to call the reader's attention to the fact that the operators $\cosh(x_1\sqrt{\frac{\partial}{\partial x_2}})$, $\frac{1}{\sqrt{\frac{\partial}{\partial x_2}}}\sinh(x_1\sqrt{\frac{\partial}{\partial x_2}})$ etc. give rise to a sheaf homomorphism between sheaves of hyperfunctions by their definition. This is a crucial point in our argument because we want to analyze the structure of (systems of) pseudo-differential equations *locally*. (See S-K-K Chapter II §5 and Chapter III §2.)

Now a linear differential operator acts on the sheaf $\mathcal{B}_M$ of hyperfunctions and the sheaf $\mathcal{A}_M$ of real analytic functions as a sheaf homomorphism by the definition, hence it acts on the quotient sheaf $\mathcal{B}_M/\mathcal{A}_M$ as a sheaf homomorphism. On the other hand we know that $\mathcal{B}_M/\mathcal{A}_M \cong \pi_*\mathcal{C}_M$ holds. Here $\mathcal{C}_M$ denotes the sheaf of microfunctions and π denotes the canonical projection from $\sqrt{-1}\,S^*M$ to M. (See the exposition of Chazarain of this issue and S-K-K Chapter I.) Hence it is natural to ask whether a linear differential operator acts the sheaf of microfunctions as a sheaf homomorphism or not. If one recalls the definition of the sheaf of microfunctions, one will easily find that the answer is affirmative.

Now, we know the following theorem due to Sato [1].

Theorem. Let $P(x,D_x)$ be a linear differential operator of finite order m defined on a real analytic manifold M. Then $P(x, D_x)$ gives rise to a sheaf isomorphism between sheaves of microfunctions outside its real characteristic variety $V=\{(x,i\eta\infty)\in \sqrt{-1}\, S^*M;\ p_m(x,i\eta)=0\}$. Here p_m denotes the principal symbol of $P(x, D_x)$. Recall that the principal symbol of a linear differential operator (of finite order m) $P(x,D_x)=\sum_{j=0}^{m} p_j(x,D_x)$ (p_j is homogeneous of order j with respect to D_x) is by definition the function obtained from $p_m(x,D_x)$ by substituting the cotangent vector η for D_x. It is well-known that the principal symbol has an intrinsic meaning independent of the choice of local coordinate system. Note that it is also well-defined on $\sqrt{-1}\, S^*M$ since it is homogeneous with respect to η.

In view of this result, it is natural to seek for a class of operators which includes the inverse of a differential operator defined outside its characteristic variety and whose element acts on the sheaf of microfunctions as a sheaf homomorphism. Such an object should be defined (locally) on $\sqrt{-1}\, S^*M$ by the above requirements. In order to find such a suitable class we use the well-known decomposition of n-dimensional δ-function into plane waves (after fixing a local coordinate system on M), i.e.

$$\delta(x-y) = \frac{(n-1)!}{(-2\pi i)^n}\int_{|\eta|=1} \frac{\omega(\eta)}{(\langle x-y,\eta\rangle+i0)^n} .$$

Now let $P(x,D_x)$ be a linear differential operator of finite order m, i.e. $P(x,D_x) = \sum_{j=0}^{m} p_j(x,D_x)$, where $p_j(x,D_x)$ is homogeneous of order j. Then we clearly have

$$(1) \quad P(x,D_x)\delta(x-y) = \sum_{j=0}^{m} \frac{(n+j-1)!}{(-1)^{n+j}(2\pi i)^n} \int \frac{p_j(x,\eta)}{(\langle x-y,\eta\rangle+i0)^{n+j}} \omega(\eta)$$

Taking account of these formula we will try to find the inverse of $P(x,D_x)$ assuming its principal symbol $p_m(x,\eta)\neq 0$ near (x^0,η^0), that is, we want to find a class of "differential operator of negative order". Then the relation $\frac{d}{dx}\log(x+i0) = \frac{1}{x+i0}$ suggests us that we should introduce some logarithmic factor as a kernel function. For this purpose we introduce a family of auxiliary functions $\Phi_\lambda(\tau)$ defined by $\frac{-1}{2\pi i}\frac{\Gamma(1+\lambda)}{(-\tau)^{\lambda+1}}$. Clearly

$\Phi_\lambda(\langle x-y,\eta\rangle+i0) = \frac{(-1)^\lambda \lambda!}{2\pi i(\langle x-y,\eta\rangle+i0)^{\lambda+1}}$ if λ is a positive integer and this is just the function used in (1). Moreover if λ is a negative integer $\Phi_\lambda(\tau)$ stands for

$$-\frac{1}{2\pi i(\lambda-1)!}\tau^{-\lambda-1}\{\log(-\tau)-(\sum_{\nu=1}^{-\lambda-1}\frac{1}{\nu}-\gamma)\}.$$

(Here γ denotes the Euler constant.) Using $\Phi_\lambda(\tau)$, we consider the following hyperfunction K in (x,y,η).

$$K = \sum_{j=-\infty}^{\infty} p_j(x,\eta)\ \Phi_j(\langle x-y,\eta\rangle + i0),$$

where $p_j(x,\eta)$ satisfies the following:

(2) $p_j(x,\eta)$ is holomorphic in a complex neighborhood U of (x^0,η^0) and homogeneous of degree j with respect to η.

(3) $$\overline{\lim_{j\to\infty}} \; j\sqrt[j]{\sup_V |p_j(x,\eta)|} = 0$$

(4) $$\overline{\lim_{j\to-\infty}} \; -\frac{1}{j} \sqrt[-j]{\sup_V |p_j(x,\eta)|} < \infty.$$

Since $\sum_{j=-\infty}^{\infty} p_j(x,\eta)\Phi_j(\langle x-y,\eta\rangle)$ converges in $W=\{(x,y,\eta);$ $|x-x^0|,|y-y^0|,|\eta-\eta^0|\ll 1,\ \mathrm{Im}\langle x-y,\eta\rangle>0\}$ by the above conditions, the hyperfunction K is well-defined as a boundary value of this holomorphic function. The above conditions (3) and (4) are natural ones of the sort so that the series $\sum p_j(x,\eta)\Phi_j(\langle x-y,\eta\rangle)$ converges in W.

It is easily verified that the microfunction defined by

(5) $$\int_{\substack{|\eta-\eta^0|\ll 1\\ |\eta|=1}} K(x,y,\eta)\omega(\eta)dy$$

has its support in $\Delta^a=\{(x,y;i(\eta,\eta')\infty);\ x=y,\ \eta=-\eta'\}$ in a neighborhood of $(x^0,\ y^0;\ i(\eta^0,-\eta^0)\infty)$ and gives rise to a sheaf homomorphism of $\mathcal{C}_M$ near $(x^0,\ i\eta^0\infty)$. (These facts follow from the theorems concerning such operations on microfunctions as integration and multiplication. See S-K-K Chapter I §2.)

The integral operator whose kernel (micro-)function is given as above will be called a pseudo-differential operator. Clearly it is a natural generalization of the notion of linear differential operators so that it should be suitable for the (local) analysis of microfunctions. Moreover Theorem 1 given below will tell us that the inverse of a linear differential operator $P(x,D_x)$ of finite order is a pseudo-differential operator as long as its principal symbol does not vanish. Thus far we have discussed pseudo-differntial operators by fixing the local coordinate system. However, it is not difficult to verify that they have an intrinsic meaning. (See S-K-K Chapter II §1.) In fact they constitute a sheaf of rings on $\sqrt{-1}\,S^*M$. Their composition rule can be expressed concretely in the following fashion.

When the kernel function of $P(x,D_x)$ has the form (5), we sometimes denote it by $\sum_{j=-\infty}^{\infty} p_j(x,D_x)$. Using this convention, the composite $R(x,D_x)=\sum_{\ell=-\infty}^{\infty} r_\ell(x,D_x)$ of two pseudo-differential operators $P(x,D_x) = \sum_{j=-\infty}^{\infty} p_j(x,D_x)$ and $Q(x,D_x) = \sum_{j=-\infty}^{\infty} q_j(x,D_x)$ can be calculated by the following rule:

$$r_\ell(x,\eta) = \sum_{\ell=j+k-|\alpha|} \frac{1}{\alpha!} D_\eta^\alpha\, p_j(x,\eta) D_x^\alpha\, q_k(x,\eta)$$

$$\left(D_\eta^\alpha = \frac{\partial^{|\alpha|}}{\partial\eta_1^{\alpha_1}\cdots\partial\eta_n^{\alpha_n}},\quad D_x^\alpha = \frac{\partial^{|\alpha|}}{\partial x_1^{\alpha_1}\cdots\,\partial x_n^{\alpha_n}}\right).$$

As in the case of differential operators, the principal symbol $p_m(x,\eta)$ of a pseudo-differential operator of finite order $P(x,D_x) = \sum_{j=-\infty}^{m} p_j(x,D_x)$ is well-defined on the cotangent bandle.

Since the verification of the growth order conditions (3) and (4) on $\{p_j(x,\eta)\}_{j=-\infty}^{\infty}$ is not so easy in general, we introduce a formal norm $N_m(P;t)$ of a pseudo-differential operator P *of finite order* m after Boutet de Monvel and Krée [1] as follows.

$$N_\ell(P;t)=\sum_{\substack{k\geq 0\\ \alpha,\beta\geq 0}}\left(\frac{2(2n)^{-k}k!}{(|\alpha|+k)!(|\beta|+k)!}\right)\sup_{\substack{(x,\eta)\in\omega\\|\eta|=1}}|D_x^\alpha D_\eta^\beta p_{\ell-k}(x,\eta)|t^{2k+|\alpha+\beta|}.$$

This is a formal power series in the auxiliary variable t with non-negative coefficients. It is easily verified by Cauch's formula that $N_\ell(P;t)<\infty$ for $0<t<\varepsilon$ is equivalent to say that estimate (4) holds. The most important property that $N_\ell(P;t)$ enjoys is the following

Proposition (Boutet de Monvel and Krée [1]) Let $P^\nu(x,D_x)$ be a pseudo-differential operator of finite order ℓ_ν. ($\nu=1,2$). Then we have $N_{\ell_1}(P^1;t)N_{\ell_2}(P^2;t)\gg N_{\ell_1+\ell_2}(P^1P^2;t)$, i.e. $N_{\ell_1}(P^1;t)N_{\ell_2}(P^2;t)$ is a majorant series of $N_{\ell_1+\ell_2}(P^1P^2;t)$.

Using the formal norm of pseudo-differential operator of finite order we can prove the following important theorem.

Theorem 1. Let $P(x,D_x)$ be a pseudo-differential operator of finite order m. Assume that its principal symbol $p_m(x,i\eta)$ never vanishes on $\Omega\subset\sqrt{-1}\,S^*M$. Then there exists a unique pseudo-differential operator $E(x,D_x)$ of finite order defined on Ω which satisfies

$$PE = EP = I.$$

Proof. Once we have proved the existence of left and right inverse of P *locally* in Ω, the assertion of the theorem can be easily verified by using the associative law for composition of pseudo-differential operators. Now we will show the local existence of left inverse. Firstly we define a pseudo-differential operator $Q_{-m}(x,D_x)$, homogeneous of order $-m$, by defining $Q_{-m}(x,\eta) = 1/p_m(x,\eta)$. Clearly $R(x,D_x) = I-Q_{-m}P$ is a pseudo-differential operator of order at most (-1). Then one can easily see by using the formal norm of pseudo-differential operators that $S = \sum_{\ell=0}^{\infty} R^{\ell}$ realy makes sense locally as a pseudo-differential operator of finite order. The pseudo-differential operator $S(x,D_x)Q_{-m}(x,D_x)$ defines a left inverse of $P(x,D_x)$ locally, since $I = S(I-R) = SQ_{-m}P$ holds by the definition.

Theorem 1 implies that we have to investigate the behavior of the microfunction solutions of the pseudo-differential equation of finite order $P(x,D_x)u=0$ only on the characteristic variety $V=\{(x,i\eta\infty)\in\sqrt{-1}S^*M;\ p_m(x,i\eta)=0\}$, since outside V $P(x,D_x)$ gives rise to a sheaf isomorphism.

In passing the characteristic variety V enjoys many geometrically interesting properties as a subvariety of $\sqrt{-1}\,S^*M$, which is a contact manifold. For example, V will be generated by the bicharacteristic strips attached to $p_m(x,i\eta)$ if V is real and regular. Hence here comes the following question naturally:

To what extent can the "geometry" —— the characteristic variety —— control the "analysis" —— the structure of microfunction solutions —— ? In other words, to what extent can the commutative objects control the non-commutative objects?

The situation is surprisingly plain: Generically, i.e. at the regular point of V, V completely determines the local structure of the pseudo-differential equation, *à fortiori*, the local structure of microfunction solutions. This is also the case for the general overdetermined systems under some moderate algebraic condition on the system (i.e., for the purely dimensional systems). (See S-K-K Chapter II Definition 5.3.6 as for the definition of purely dimensional systems. Note that any system with one unknown is purely dimensional.)

This is the main result obtained in S-K-K Chapter II and Chapter III. See Theorem 2.4.1 in Chapter III and Theorem 5.3.7 in Chapter II especially.

The proof of the above statement is very long and needs a great deal of preparation. However, its basic idea is straightforward and we present it here rather schematically. See S-K-K for details.

The proof will be divided into two steps.

The first one deals with the geometrical investigation of the characteristic variety V, which is a subvariety of pure imaginary contact manifold $\sqrt{-1}\, S^*M$. Here the important point is the following observation originally due to Maslov [1] and Egorov [1] (see also Hörmander [1] and S-K-K Chapter II § 3.3 and § 4.3): We can associate a transformation of pseudo-differential operators to any contact transformation so that the principal symbol of the operators are transformed according to the contact transformation. (For the precise statement, see S-K-K Chapter II Theorem 3.3.3 and /or Theorem 4.3.1 for example.)

Therefore obtaining the standard form of V under the (real) contact transformation reduces the problem to the case where the principal symbol of the pseudo-differential operator takes such a simple form as either η_1, $\eta_1+i\eta_2$ or $\eta_1+ix_1\eta_2$. (In the case of over-determined system, their combination appears.) This topic is discussed in S-K-K Chapter III §2. See also Sato, Kawai and Kashiwara [2].

Here it is essential that we consider all problems locally on $\sqrt{-1}\,S^*M$, not locally on M. Here appears one of the greatest advantages of employing the notion of microfunctions.

The next step is to deal with the lower order terms once we have obtained the standard form of the principal symbol. Here the use of pseudo-differential operators of infinite order is very crucial as is easily seen from the example discussed at the beginning of this talk. In fact we cannot transform the equation

$\left(\frac{\partial^2}{\partial x_1^2} - \frac{\partial}{\partial x_2}\right) u = 0$ to the equation $\frac{\partial^2}{\partial x_1^2} u = 0$ if we

use only pseudo-differential operator of finite order. The fact that we need not pay attention to the lower order terms at the generic point of V can be established only by the aid of pseudo-differential operators of infinite order. This is the result established in S-K-K Chapter II §5. Note that we have used pseudo-differential operators of infinite order only as an auxiliary tool in transforming the complicated equation to a simpler one. Plainly speaking, the theory of pseudo-differential equations of infinite order is quite beyound us.

This is the main idea in establishing the desired result: "Commutative object" completely determines the "non-commutative object" at the regular point of the characteristic variety V.

Thus far we have mainly discussed the pseudo-differential operators from the view point that they act on the sheaf of microfunctions as sheaf homomorphisms. However, the notion of pseudo-differential operators itself is not inherent to the real manifold. In fact it can be defined on the complex manifold, though the sheaf of microfunctions cannot be defined there. Hence we can talk about the structure of pseudo-differential equations even in the complex domain. This is the topic discussed in S-K-K Chapter II.

In order to make this point clearer we call the reader's attention to the following fact: Pseudo-differential operators cannot exhaust all integral operators that give rise to sheaf homomorphisms on the sheaf of microfunctions. In fact any operator whose kernel (micro-) function has its support in $\Delta_M^a=\{(x,x';$ $i(\eta,\eta')\infty)\in\sqrt{-1}\,S^*(M\times M);\ x=x',\ \eta=-\eta'\}$ does enjoy such a property. Of course the class of all integral operators that give rise to sheaf homomorphisms is too wide and is not suitable for concrete calculation, which is possible in the case of pseudo-differential operators. However, the employment of such operators — called micro-local operators in S-K-K — is still very important in some points. In fact, the celebrated counter-example of Lewy [1] is best illustrated by employing micro-local operators and it is also crucial in investigating the structure of microfunction solutions of general (system of) pseudo-differential equations. See S-K-K Chapter III §2.3 and §2.4 for details. Here we explain only the Lewy phenomena briefly.

Let P be the Lewy operator, i.e., $\frac{\partial}{\partial \bar{z}} - iz\frac{\partial}{\partial t} = \frac{1}{\partial}(\frac{\partial}{\partial x} + i\frac{\partial}{\partial y}) - i(x+iy)\frac{\partial}{\partial t}$ and let Q be $\bar{P} = \frac{\partial}{\partial z} + i\bar{z}\frac{\partial}{\partial t}$. Here (z,t) denotes a point in $\mathbb{C} \times \mathbb{R} \cong \mathbb{R}^3$. Let us define the following kernel function K by

$$-\frac{1}{\pi^2}\frac{1}{(t-t' + i(|z|^2-|z'|^2-2z\bar{z}') + i0)^2} .$$

It is easily verified that the support of K as a microfunction in $(z,z',t,t') \in \mathbb{R}^3 \times \mathbb{R}^3$ is contained in $\Delta^a_{\mathbb{R}^3}$.

If we denote by $\mathcal{K}$ the micro-local operator defined by the kernel function K, then we have the following exact sequence on $\Omega = \{(z,t;i(\zeta,\tau)\infty) \in \sqrt{-1}\, S^*\mathbb{R}^3;\ \tau > 0\}$.

$$0 \longrightarrow \mathcal{C} \xrightarrow{\ Q\ } \mathcal{C} \xrightarrow{\ \mathcal{K}\ } \mathcal{C} \xrightarrow{\ P\ } \mathcal{C} \longrightarrow 0$$

This exact sequence tells us that the Lewy operator P is surjective on Ω, while its complex conjugate Q is injective on Ω. Moreover the range of Q and the null-space of P are characterized by the micro-local operator K. Thus we have found the concrete obstruction against the solvability of the equation $Qu = f$. It is also easily seen in the same way that we can find the analogous exact sequence in $\Omega^- = \{(z,t;i(\zeta,\tau)\infty);\ \tau < 0\}$ by replacing the roles of P and Q, that is, P is injective but not surjective there and that Q is surjective but not injective there. These observations have given the lucid explanation for the Lewy phenoma and at the same time they show clearly the advantage of employing the theory of microfunctions in the theory of linear differential equations.

References

Boutet de Monuel, L. and P. Krée: [1] Pseudo-differential operators and Gevrey classes, Ann. Inst. Fourier, 17(1967) 295-323.

Egorov, Yu. V.: [1] On canonical transformations of pseudo-differential operators, Uspehi Mat. Nauk, 25(1969) 235-236.

Hörmander, L.: [1] Fourier integral operators, I, Acta Math. 127(1971) 79-183.

Lewy, H.: [1] An example of a smooth linear partial differential equation without solution, Ann. of Math. 66(1957) 155-158.

Maslov, V.: [1] Theory of Perturbation and Asymptotic Method, Moscow State Univ. 1965(Russian, also translated into French by Lascoux and Seneor (Dunod-Ga thier-Villars, 1972).

Sato, M: [1] Hyperfunctions and partial differential equations, Proc. Intern. Conf. on Functional Analysis and Related Topics, Univ. of Tokyo Press, 1969, pp.91-94.

Sato, M., T. Kawai and M. Kashiwara: [1] Microfunctions and pseudo-differential equations, Proc. Katata Conf., Lecture Notes in Math. No.287, Springer, 1973, pp.263-529.

______________ : [2] On the structure of single linear pseudo-differential equations, Proc. Japan Acad. 48(1972) 643-646.

MICRO-HYPERBOLIC PSEUDO-DIFFERENTIAL OPERATORS

Masaki KASHIWARA and Takahiro KAWAI

Research Institute for
Mathematical Sciences,
Kyoto University

and

Département de Mathématiques, Nice

Study of (fundamental solutions of) hyperbolic differential equations has a long history. See for example Riemann [1], Hadamard [1], Courant-Hilbert [1] and references cited in Courant-Hilbert [1], [2]. In such a long history the works of Petrowsky [1] and Gårding [1] are clearly outstanding milestones from the view point of the general theory of differential equations because of the generality of their results. Leray [1] has also influenced much the later development of the theory of hyperbolic differential equations by establishing the existence and uniqueness theorems of the solutions in an elegant and far-reaching way. See also Friedrichs-Lewy [1] and Friedrichs [1]. On the other hand Hörmander [1], [2] gave good existence and uniqueness theorems for real operators of principal type along these lines. One of the reasons for the success of Hörmander seems to us to be the fact that such operators are micro-hyperbolic. The notion of micro-hyperbolicity was introduced in Kashiwara-Kawai [1] in full generality. Hereafter Kashiwara-Kawai [1] will be often referred to as K-K for short. In the case of linear differential operators with constant coefficients the same notion was introduced by Andersson [1] influenced by the work of Atiyah-Bott-Gårding [1]. (See Kawai [2] and Gårding [2].) In most of the above quoted papers the fundamental solution plays its essential role. As for the construction and investigation of the fundamental solutions for hyperbolic operators or real operators of principal type, we refer to Courant-Lax [1], Lax [1], Leray [1], [2], Ludwig [1], Mizohata [1], Hörmander [3], Kawai [1] and Duistermaat-Hörmander [1]. Note that these works assume the regularity of characteristic variety of the operator. In order to treat the operators whose characteristic variety is not simple, the employment of hyperfunctions is very crucial as is shown by Bony-Schapira [1], [2]. (See Mizohata [2], where a necessary condition for hyperbolicity is discussed. See also Leray-Ohya [1], Mizohata-Ohya [1], Chazarain [1] and the references cited there about the operators with constant multiple characteristics.)

The purpose of this report is to explain in a sketchy way the idea of Kashiwara-Kawai [1], whose results cover all the above quoted (local) existence and uniqueness theorems as long as the operators under consideration have analytic coefficients. Note that the theory developed in Kashiwara-Kawai [1] is also related to Egorov [1], [2], Nirenberg-Treves [1] and Treves [2].

(Theorem 3 in the below.)

The topics of this report are completely restricted to the existence and uniqueness of the (fundamental) solutions of linear (pseudo-) differential equations and other topics of hyperbolic equations are not discussed here, though some of them are important and also expected to be closely related to the topics discussed here, e.g. hyperbolic mixed problems. (As for such problems we refer to the exposition of Chazarain [2] for example.)

Now we will sketch the idea of the proof of the existence of fundamental solutions for partially micro-hyperbolic pseudo-differential operator $P(x,D_x)$.

A pseudo-differential operator $P(x,D_x)$ is said to be partially micro-hyperbolic at $(x^0, i\xi^0) \in \sqrt{-1}S^*M$ with respect to the direction $\langle\vartheta,dx\rangle+\langle\rho,d\xi\rangle$ if $p_m(x+i\varepsilon\rho, i\xi+\varepsilon\rho)\neq 0$ for every (x,ξ) sufficiently close to (x^0, ξ^0) and for $0<\varepsilon\ll 1$. (See K-K §1 for the precise definition. See Gårding [2], [3] also.) Then using the "quantized" contact transformation (Sato-Kawai-Kashiwara [1] Chapter II §3.3) we can easily reduce the problem to the case where P has a matrix form $D_{x_1}-A(x,D')$ so that A is a square matrix of pseudo-differential operators of order ≤ 1 which commute with x_1 and that all eigenvalues of its principal symbol $A_1(x, i\xi')$ have non-negative real part. (The above quoted report of Sato, Kawai and Kashiwara will be referred to S-K-K [1] hereafter.)

The first step in our arguments is to construct formally a solution $R(x,D')=\Sigma\, a_\alpha(x)D'^\alpha$ as an infinite sum of pseudo-differential operators so that $PR=0$ and that $R|_{x_1=0}=1$. This part of the proof is not difficult to perform. In fact we need not use the assumption of partial micro-hyperbolicity of P at this stage. Cf. Treves [1]. What we need here is that $\{x_1=0\}$ is non-characteristic with respect to P. (See Proposition 2.2 in K-K §2.) The essential difficulty comes in at the next step. Generally the existence domain of R is so small that R cannot be endowed with the meaning as a kernel microfunction of a fundamental solution of P. In order to overcome this difficulty we rewrite the equation $PR=0$ by using the "defining function" of P and R. (Lemma 4.1 in K-K §4.) Then we try to extend the domain of definition of the defining function G of R by the partial micro-hyperbolicity of P. In extending the domain of definition the following Lemma 1 is crucial. Note that the partial micro-hyperbolicity of $P=D_{x_1}-A(x,D')$ at $(x_1,x';i(\xi_1,\xi')\infty)=(0,x^{0'};i(\xi_1,\xi^{0'})\infty)=(0;i(\xi_1,0,\ldots,0,1)\infty)$ for every real ξ_1 with respect to the direction x_1 implies that

$$g(x_1,z',\zeta_1,\zeta') = \det(\zeta_1 - A_1(x_1,z',\zeta'))$$

never vanishes on

$$\{(x_1,z',\zeta_1,\zeta')\in \mathbb{R}\times\mathbb{C}^{n-1}\times\mathbb{C}\times\mathbb{C}^n;\ 0\leq x_1\leq\delta,$$
$$|z'|<\delta,\ |(\zeta_2,\ldots,\zeta_{n-1})|=|\zeta''|<\delta|\zeta_n|,\ -\mathrm{Im}(\zeta_1/\zeta_n)$$
$$> M(|y| + \sum_{\nu=2}^{n-1}|\mathrm{Im}(\zeta_\nu/\zeta_n)|.\}$$

Here the assumption of the partial micro-hyperbolicity of P for every real ξ_1 is not restrictive in application because we can easily localize the problem with respect to ξ_1 by the preparation theorem of Weierstrass for pseudo-differential operators (S-K-K [1] Chapter II §2.2.) See the arguments in the proof of Theorem 5.2 in K-K §5 for details.

After the above observation concerning the implication of partial micro-hyperbolicity, we state the following lemma.

Lemma 1. Let $\varphi(x_1,z',\bar{z}')$ be a positive valued real analytic function defined on $U=\{(x_1,z')=(x_1,x'+iy')\ ;\ 0<x_1<\delta_1,$ $|x'|<\delta_2,\ |y'|<\delta_3$ with $\delta_2{}^2+\delta_3{}^2<\delta_1{}^2\}$. Assume that φ satisfies the following:

$$(1)\qquad \frac{\partial\varphi}{\partial x_1} > M\left(|y'| + \varphi + \sum_{\nu=2}^{n-1}\left|\frac{\partial\varphi}{\partial x_\nu}\right|\right)$$

$$(2)\qquad \sum_{\nu=2}^{n-1}\left|\frac{\partial\varphi}{\partial z_\nu}\right| < \frac{\delta}{2} \quad \text{on } U.$$

Suppose that G can be extended to $V=\{z;\ 0<x_1<\delta_1,$ $|z'|<\delta_2,\ y_n>\varphi(x_1,z',\bar{z}')\}$. Then G can be extended to a holomorphic function defined on an open set V' which contains

$$\{z;\ 0 < x_1 < \delta_1,\quad |z'| < \delta_2,\quad y_n \geq \varphi(x_1,z',\bar{z}')\}.$$

Once we have proved this lemma, we can easily prove the following Theorem 2 by a suitable choice of φ, while the proof of the above lemma is reduced to the invertibility of elliptic pseudo-differential operator. (See S-K-K [1] Chapter II §2.1 Theroem 2.1.1. See also the exposition of Kawai of this issue.)

Theorem 2. There exist $\delta_0>0$ and M_1 such that $G(x_1,z')$ is holomorphic

$$\{(x_1,z')\in \mathbb{R}\times\mathbb{C}^n;\ \ 0<x_1<\delta_0,\ \ |z'|<\delta_0,$$
$$\operatorname{Im} z_n > M_1x_1(\sum_{\nu=2}^{n-1}|\operatorname{Im} z_\nu|)\}$$

As for the details of the proof of Lemma 1 and Theorem 2 we refer to K-K §4.

Now Theorem 2 allows us to define the boundary value $w(x)$ of a hyperfunction $G^+(x_1,z')=Y(x_1)G(x_1,z')$ with holomorphic parameters z' defined on

$$\{(x_1,z');\ \ |x_1|,\ |z'|<\delta,\ \ \operatorname{Im} z_n > M|x_1|(\sum_{\nu=2}^{n-1}|\operatorname{Im} z_\nu|)\}.$$

(See S-K-K Chapter I §3.2 about the notion of taking the boundary value of hyperfunctions with holomorphic parameters.)

It is readily verified that the singular spectrum $u(x)$ of $w(x)$ satisfies $Pu=\delta(x)$ and that $\operatorname{Supp} u\subset\{(x;\ i\xi\infty);\ x_1\geqq 0,$ $|\xi_\nu|\leqq Mx_1|\xi_n|\ (\nu=2,\ldots,n-1),\ |x_n|\leqq \nu x_1\}$. Using this fact one can easily show the existence of fundamental solution of the partially micro-hyperbolic pseudo-differential operator P. (See Theorem 5.2 in K-K §5.)

Once one gets a fundamental solution, it is easy to show the existence or (propagation of) regularity of solutions. The results are listed up in §6 of K-K and we omit the details here. However, we would like to touch the following theorem without proof. This theorem will show why we have treated the *partially* micro-hyperbolic operators, not the micro-hyperbolic operators. In fact, $D_{x_1}+ix_1^{2k}D_{x_2}$ (Mizohata [3]), the easiest and most typical example that can be covered by Theorem 3, is not micro-hyperbolic, though it is partially micro-hyperbolic. (See Sato-Kawai-Kashiwara [2] also.)

Theorem 3. Assume that the real characteristic variety V of $P(x,D_x)$ is defined by $a(x,\eta)+\sqrt{-1}b(x,\eta)=0$ where $(\sqrt{-1})^{-m}a(x,\sqrt{-1}\eta)$ and $(\sqrt{-1})^{-m}b(x,\sqrt{-1}\eta)$ are real for $(x,\sqrt{-1}\eta)$ in $\sqrt{-1}S^*M$ near $x_0^*=(x_0,\sqrt{-1}\eta_0)$ and that $\text{grad}_{(x,\eta)}a(x,\eta)$ and ω are linearly independent there. (Here m denote the degree of a and b with respect to η.) Assume further that $(\sqrt{-1})^{-m}b(x,\sqrt{-1}\eta)$ is positive (or negative) on each real bicharacteristic strip of $(\sqrt{-1})^{-m}a(x,\sqrt{-1}\eta)$ and not identically zero there. Then $P(x,D_x)$ has an inverse in the ring of micro-local operators.

We refer to S-K-K [1] Chapter I §2.5 and the exposition of Kawai of this issue about the notion of micro-local operators. Note that the above theorem implies not only the micro-local solvability of the equation $Pu=f$ but also the "micro-local" analytic-hypoellipticity of P. We also note that a more general result is given in K-K. (Theorem 6.6 in §6.)

At the end of this exposition the speakers would like to lay stress on the following point as a summary:

The employment of hyperfunctions and microfunctions has made the theory of linear hyperbolic differential equations very lucid and thrown the light to the nature of a class of hypoelliptic operators from the view-point of "hyperbolicity." The essential idea in showing these is "taking the boundary value of pseudo-differential operators defined in the complex domain." In fact $P(x,D_x)$ is invertible when $p_m(x,\eta)\neq 0$ and the partial micro-hyperbolicity of $P(x,D_x)$ means the invertibility of P on a conical set which is tangent to the real axis. Therefore what we have done may be summarized as a justification of the procedure of "taking the boundary value of pseudo-differential operators."

References

Andersson, K. G.: [1] Propagation of analyticity of solutions of partial differential equations with constant coefficients, Ark. Mat., 8 (1971) 277-302.

Atiyah, M. F., R. Bott and L. Gårding: [1] Lacunas for hyperbolic differential operators with constant coefficients I., Acta Math., 124 (1970) 109-189.

Bony, J. M. et P. Schapira: [1] Problème de Cauchy, existence et prolongement pour les hyperfonctions solutions d'équations hyperboliques non strictes, C. R. Acad. Sci. Paris, 274 (1972) 188-191.

_____: [2] Solutions hyperfonctions du problème de Cauchy, Hyperfunctions and Pseudo-differential Equations, Lecture Notes in Mathematics No. 287, Springer, Berlin-Heidelberg-New York, 1973, pp. 82-98.

Chazarain, J.: [1] Opérateurs hyperboliques à caracteristiques de multiplicité constante, to appear.

_____: [2] An article which will appear in Sém. Bourbaki (1972/1973).

Courant, R. u. D. Hilbert: [1] Methoden der Mathematischen Physik, II, Springer, Berlin, 1937.

_____: [2] Methods of Mathematical Physics, II, Interscience, New York, 1962.

Courant, R. and P. D. Lax: [1] The propagation of discontinuities in wave motion, Proc. Nat. Acad. Sci. U.S.A., 42 (1956) 872-876.

Duistermaat, J. and L. Hörmander: [1] Fourier integral operators II, Acta Math., 128 (1971) 183-269.

Egorov, Yu. V.: [1] Conditions for the solvability of pseudo-differential operators, Dokl. Akad. Nauk USSR, 187 (1969) 1232-1234. (In Russian.)

_____: [2] On subelliptic pseudo-differential operators, Dokl. Akad. Nauk USSR, 188 (1969) 20-22. (In Russian.)

Friedrichs, K. O.: [1] Symmetric hyperbolic system of linear differential equations, Comm. Pure Appl. Math., 7 (1954) 345-392.

Friedrichs, K. O. u. H. Lewy: [1] Über die Eindeutigkeit und das Abhängigkeitsgebiet der Lösungen beim Anfangsproblem linearer hyperbolischer Differentialgleichungen, Math. Ann., 98 (1928) 177-195.

Gårding, L.: [1] Linear hyperbolic partial differential equations with constant coefficients, Acta Math., 85 (1950) 1-62.

_____: [2] Local hyperbolicity, Israel J. Math., 13 (1972) 65-81.

_____: [3] A note which will appear in Israel J. Math. as a supplement to Gårding [2].

Hadamard, J.: [1] Lectures on Cauchy Problem in Linear Partial Differential Equations, Yale Univ. Press, 1923. Reprinted by Dover, New York.

Hörmander, L.: [1] On the theory of general partial differential operators, Acta Math., 94 (1955) 161-184.

_____: [2] Linear Partial Differential Operators, Springer, Berlin-Heidelberg-New York, 1963.

_____: [3] On the singularities of solutions of partial differential equations, Proc. Int. Conf. Functional Analysis and Related Topics, Univ. of Tokyo Press, Tokyo, 1970, pp.31-40.

Kashiwara, M. and T. Kawai: [1] Micro-hyperbolic pseudo-differential operators I, to appear.

Kawai, T.: [1] Construction of local elementary solutions for linear partial differential operators with real analytic coefficients (I)— The case with real principal symbols —, Publ. RIMS, Kyoto Univ., 7 (1971) 363-397.

_____: [2] On the global existence of real analytic solutions of linear differential equations (I), J. Math. Soc. Japan, 24 (1972) 481-517.

Lax, P. D.: [1] Asymptotic solutions of oscillatory initial value problems, Duke Math. J., 24 (1957) 627-646.

Leray, J.: [1] Hyperbolic Differential Equations, The Institute for Advanced Study, Princeton, 1952.

_____: [2] Un prolongement de la transformation de Laplace qui transforme la solution unitaire d'un opérateur hyperbolique en sa solution élémentaire, Bull. Soc. Math. Fr., 90 (1962) 39-156.

Leray, J. and Y. Ohya: [1] Équations et systèmes non-linéaires, hyperboliques non-stricts, Math. Ann., 170 (1967) 167-205.

Ludwig, D.: [1] Exact and asymptotic solutions of the Cauchy problem, Comm. Pure Appl. Math., 13 (1960) 473-508.

Mizohata, S.: [1] Analyticity of solutions of hyperbolic systems with analytic coefficients, Comm. Pure Appl. Math., 14 (1961) 547-559.

_____: [2] Some remarks on the Cauchy problem, J. Math. Kyoto Univ., 1 (1961) 109-127.

_____: [3] Solutions nulles et solutions non-analytiques, ibid. 271-302.

Mizohata, S. and Y. Ohya: [1] Sur la condition d'hyperbolicité pour les équations a characteristiques multiples, II, Japanese J. Math., 40 (1971) 63-104.

Nirenberg, L. and F. Treves: [1] On local solvability of linear partial differential equations —— Part II. Sufficient conditions, Comm. Pure Appl. Math., 23 (1970) 459-510.

Petrowsky, I. G.: [1] Über das Cauchysche Problem für Systeme von partiellen Differentialgleichungen, Mat. Sbornik, 44 (1937), 815-868.

Riemann, B.: [1] Über die Fortpflanzung ebener Luftwellen von endlicher Schwingusweite, Mathematische Werke, Dover, New York, 1953, pp. 156-175.

Sato, M., T. Kawai and M. Kashiwara: [1] (Refered to as S-K-K[1]) Microfunctions and pseudo-differential equations,

Hyperfunctions and Pseudo-differential Equations, Lecture Notes in Mathematics. No. 287, Springer, Berlin-Heidelberg-New York, 1973, pp. 265-529.

_____: [2] On the structure of single linear pseudo-differential equations, Proc. Japan Acad., 48 (1972), 643-646.

Treves, F.: [1] Ovcyannikov Theorem and Hyperdifferential Operators, I. M. P. A., Rio-de-Janeiro (Brasil), 1969.

_____: [2] Analytic-hypoelliptic partial differential equations of principal type, Comm. Pure Appl. Math., 24 (1971) 537-570.

MICROANALYTICITE DE LA MATRICE S

Frédéric PHAM

Département de Mathématiques, NICE

La physique des particules élémentaires s'intéresse à des "processus de collision", qu'on a l'habitude de noter de façon analogue à des réactions chimiques, p.ex.

$$\pi^- + p \longrightarrow K^\circ + K^\circ + \Xi^\circ \quad ,$$

type de formule que nous résumerons par la notation générique*)

$$I \longrightarrow J$$

ou plus simplement (IJ) .

Il sera aussi commode de noter un tel processus par un graphe, p. ex.

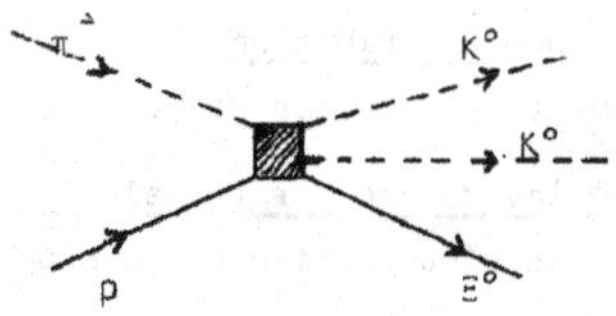

Si l'on ne tient compte que des "interactions fortes" (négligeant les interactions électromagnétiques et autres interactions "faibles"), et si l'on ne considère que des processus de collisions entre particules stables ("stables" s'entend relativement aux interactions fortes), on peut admettre que longtemps avant (resp.longtemps après) la collision, les particules entrantes (resp.sortantes) sont des particules

*) Il sera commode d'interpréter les lettres I, J de façon "ensembliste" : dans l'exemple ci-dessus, I = {méson π^-, proton} , J = {1er méson K°, 2e méson K°, hyperon Ξ°} ; la numérotation des particules de même type (les deux mésons K° dans cet exemple) n'est qu'un artifice mathématique, absolument vide de sens physique en vertu du principe d'indiscernabilité des particules.

libres (car les interactions fortes sont à courte portée). Rappelons que l'état cinématique d'une particule libre peut être caractérisé (en oubliant le nombre quantique discret qu'est le "spin") par son quadri-vecteur d'impulsion-énergie,

$$p = (p_{(o)}, \vec{p}) \in \mathbb{R} \times \mathbb{R}^3 = \mathbb{R}^4$$

dont la composante $p_{(o)} > 0$ représente l'énergie, et dont le carré scalaire dans la métrique de Minkowski est égal au carré de la masse de la particule

$$p^2 \equiv p_{(o)}^2 - \vec{p}^2 = m^2$$

(les unités sont choisies de façon que la vitesse de la lumière vale 1). Nous noterons M, et appellerons couche de masse de la particule, la nappe d'hyperboloïde ainsi définie :

$$M = \{p \in \mathbb{R}^4 \mid p^2 = m^2 , p_{(o)} > 0\} \quad .$$

En vertu du "principe de superposition" (principe fondamental de toute la physique quantique), toute l'information qu'il est possible d'extraire d'expériences de collision $I \rightarrow J$ est contenue dans ce qu'on appelle l'élément de matrice S du processus $I \longrightarrow J$: c'est une distribution S_{IJ}, à valeurs complexes, sur la variété $\prod_{i \in I \sqcup J} M_i$ (produit des couches de masse de toutes les particules en jeu). A partir du principe d'invariance par translation des lois de la physique, il est facile de montrer que cette distribution a son support dans la sous-variété*)

$$M_{(IJ)} = \{(p_i) \in \prod_{i \in I \sqcup J} M_i \mid \sum_{i \in I} p_i = \sum_{j \in J} p_j\}$$

(c'est la loi de "conservation de l'impulsion-énergie") ; plus précisément, on montre que S_{IJ} est de la forme

$$S_{IJ} = \delta(\sum_{i \in I} p_i - \sum_{j \in J} p_j) \, S_{(IJ)}$$

*) Comme les autres variétés algébriques que nous introduirons par la suite, cette "variété" est lisse pour des valeurs génériques des masses, mais elle acquiert des singularités pour des valeurs des masses qui, bien que "particulières", se rencontrent effectivement en physique - p.ex. chaque fois que les masses des particules sortantes sont égales à celles des particules entrantes, ce qui est le cas des "processus élastiques" (processus $I \rightarrow I$). Ces cas "particuliers" conduisent à des difficultés techniques que - faute de les avoir résolues - j'exorciserai par l'incantation suivante : "on se ramène au cas générique par perturbation des masses".

où δ est la distribution de Dirac (ou plutôt le produit de 4 distributions de Dirac, une par composante de quadri-vecteur), et $S_{(IJ)}$ est une distribution sur la sous-variété $M_{(IJ)}$, distribution appelée <u>amplitude de diffusion</u> du processus $I \to J$.

Dans ce qui suit, on va énoncer des <u>"hypothèses de microanalyticité"</u> des amplitudes de diffusion. On ne cherchera pas à justifier physiquement ces hypothèses (ceci sera fait dans l'exposé de Iagolnitzer), mais on tâchera de les présenter sous une forme mathématique telle que l'interprétation physique apparaisse "en filigrane".

Il ressort des progrès récents de la <u>"théorie axiomatique de la matrice</u> S" que ces hypothèses de microanalyticité (ou les hypothèses équivalentes de "macrocausalité" dont parlera Iagolnitzer) peuvent être prises comme axiomes fondamentaux de cette théorie. En théorie axiomatique des champs, où la matrice S n'est pas l'ingrédient de départ mais un sous-produit, ces hypothèses "devraient" pouvoir être démontrées comme conséquence des axiomes (mais tout le travail reste à faire !).

I. MICROANALYTICITE DES AMPLITUDES DE DIFFUSION.

1.1. Espace cotangent à la variété $M_{(IJ)}$.

L'espace de Minkowski des quadrivecteurs d'impulsion énergie peut être mis en dualité avec l'espace de Minkowski des translations d'espace-temps : si p est un quadrivecteur d'impulsion-énergie et u une translation d'espace-temps, le produit scalaire $p.u$ (dans la métrique de Minkowski) est un invariant de Lorentz qui a la dimension d'une action (produit d'une énergie par un temps) ; or il existe dans la nature une constante fondamentale qui a la dimension d'une action, la "constante de Planck" ; $p.u$ divisé par la constante de Planck est donc un scalaire au sens mathématique du terme.

Par cette dualité, il est évident que l'espace cotangent à la couche de masse M en un point $p \in M$ s'identifie au quotient de l'espace des translations d'espace-temps par le sous-espace (à une dimension) des translations parallèles à p .Autrement dit, si l'on appelle "trajectoire libre" d'impulsion-énergie p toute droite de l'espace-temps parallèle à p , l'espace cotangent à M en p peut s'interpréter comme l'espace vectoriel des translations des trajectoires libres d'impulsion-énergie p .

De façon analogue, on vérifiera que l'espace cotangent $T^*_{(p)}M_{(IJ)}$ à la variété $M_{(IJ)}$ en un point $(p) = (p_i)_{i\in I\amalg J}$ de cette variété peut s'interpréter comme l'espace vectoriel des "translations relatives" d'une famille de trajectoires libres d'impulsions-énergies p_i $(i \in I\amalg J)$, c.à.d. les translations modulo les translations d'ensemble*) .

1.2. Configuration élémentaire associée à un point $(p) \in M_{(IJ)}$.

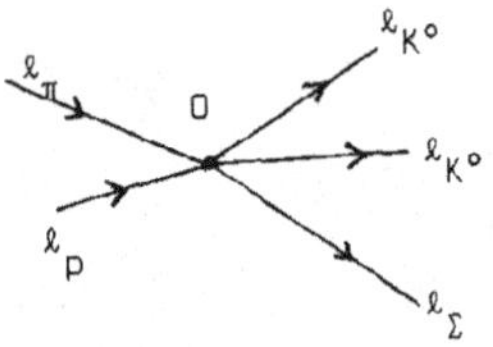

--------> (axe des temps)

Nous appellerons configuration élémentaire associée à un processus (IJ) toute famille de demi-droites orientées $(\ell_i)_{i\in I\amalg J}$ de l'espace-temps, incidentes

---

*) Pour arriver à cette interprétation, il faut définir la dualité dans l'espace $(\mathbb{R}^4)^{I\amalg J}$ par la formule : $(p).(u) = \sum_{i\in I} p_i.u_i - \sum_{j\in J} p_j.u_j$. Evidemment, faute d'entrer dans des justifications physiques détaillées, notre seul argument pour justifier ce choix est la simplicité du résultat.

à un même point O , et telles que

1°) chaque ℓ_i est de genre temps, et orientée dans le sens du temps ;

2°) ℓ_i a le point O comme extrêmité ou comme origine selon que $i \in I$ ou J;

3°) si l'on désigne par p_i l'unique quadrivecteur parallèle à ℓ_i tel que $p_i^2 = m_i^2$ (m_i = masse de la particule i) , on a

$$\sum_{i \in I} p_i = \sum_{j \in J} p_j \quad .$$

Il est clair par définition que les configurations élémentaires associées à un processus (IJ) sont, modulo le choix du point O , en correspondance bijective avec les points $(p) \in M_{(IJ)}$.

1.3. Configuration causale.

Nous appelons configuration causale toute famille finie de trajectoires (dans l'espace-temps) de particules suivant les lois de la cinématique relativiste classique et interagissant ponctuellement : l'exemple le plus simple de configuration causale est la "configuration élémentaire" que nous venons de définir ; plus généralement, une configuration causale est une famille de trajectoires rectilignes (droites, demi-droites, ou segments de droites, du genre temps et orientées dans le sens du temps) telle que si l'on appelle "sommet" de la configuration tout point où l'une des trajectoires commence ou se termine, la configuration coïncide au voisinage de chacun de ses sommets avec une configuration élémentaire.

Les trajectoires bornées (segments de droites) de la configuration sont appelées "lignes internes" , les autres sont les "lignes externes".

1.4. Covecteur causal.

Soit (u) un vecteur cotangent à $M_{(IJ)}$ en un point $(p) \in M_{(IJ)}$. Un tel covecteur peut être représenté par une famille de translations $(u_i \in \mathbb{R}^4)_{i \in I \amalg J}$. En faisant agir les translations u_i sur les lignes ℓ_i de la configuration élémentaire associée à (p) , on obtient une famille de demi-droites $u_i(\ell_i)$. Il peut se faire que cette famille de demi-droites coïncide avec la famille des lignes externes d'une configuration causale C (figure ci-dessous).

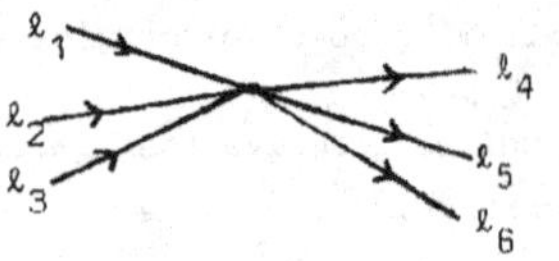

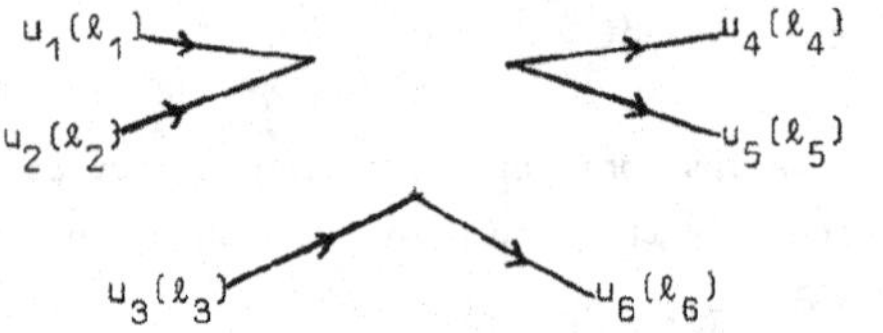

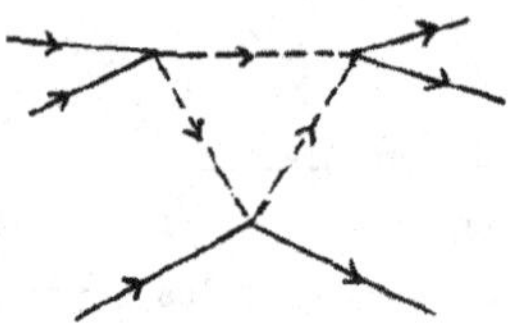

Configuration causale C

Si c'est le cas (pour un choix convenable du représentant $(u_i \in \mathbb{R}^4)$ du covecteur (u)), nous dirons que le covecteur (u) et la configuration causale C sont associés.

Nous appellerons covecteur causal tout covecteur associé à une configuration causale.

On observera que, puisque la notion de configuration causale est stable par dilatation, le caractère causal ou non d'un covecteur se conserve quand on multiplie ce covecteur par un scalaire positif, autrement dit, ne dépend que de la direction de ce covecteur.

On notera $S^*M_{(IJ)}$ le fibré (sur $M_{(IJ)}$) de toutes les directions de covecteurs (c.à.d. le fibré en sphères associé au fibré cotangent $T^*M_{(IJ)}$) .

Tout ce qui précède nous permet d'énoncer l'

HYPOTHESE DE MICROANALYTICITE :

L'amplitude de diffusion $S_{(IJ)}$ est microanalytique dans la direction de tout covecteur $(u) \in T^*M_{(IJ)}$, à l'exception des covecteurs causaux.

Cette hypothèse permet en principe de délimiter le support spectral de $S_{(IJ)}$ (dans $S^*M_{(IJ)}$) à partir de la seule donnée du "spectre de masse" des particules existant dans la nature, c.à.d. du sous-ensemble $\mathfrak{S} \subset \mathbb{R}^+$ formé de toutes les masses de particules existantes. On suppose généralement que ce spectre de masse est discret et non adhérent à zéro (hypothèse d'une "masse minimale"). Dans ces conditions, on a le résultat suivant (H.P. Stapp) ;

1.5. Convenons que deux configurations causales sont de même type s'il existe un homéomorphisme de l'espace-temps $\mathbb{R}^4$ qui transforme l'une en l'autre de telle façon que chaque ligne soit transformée en une ligne porteuse de la même masse (si plusieurs particules suivent la même trajectoire, la masse portée est la somme des masses). Alors, l'ensemble des types de configurations causales est localement fini sur $M_{(IJ)}$ (ou sur $S^*M_{(IJ)}$).

En particulier un covecteur causal ne peut correspondre qu'à un nombre fini de types de configurations causales. S'il ne correspond qu'à un seul type, je dirai que c'est un covecteur causal simple. La question suivante me semble naturelle : "presque tout" covecteur causal est-il simple ? (je ne connais pas la réponse).

2. SINGULARITES DE LANDAU.

2.0. Oublions la structure métrique (dans l'espace-temps) d'une configuration causale pour n'en retenir que la structure combinatoire : on obtient le "graphe de diffusion multiple" associé à la configuration causale ; c'est un graphe (abstrait) orienté, sans circuits, dont chaque ligne interne ou externe porte le nom d'une particule et dont chaque sommet a pour étoile le graphe d'un processus de collision. Nous dirons qu'un graphe de diffusion multiple est associé à un covecteur $(u) \in T^*M_{(ij)}$ s'il est associé à une configuration causale associée à (u).

REMARQUE : Le lecteur peut se demander quelle différence il y a entre la donnée d'un "graphe de diffusion multiple" et la donnée d'un "type de configuration causale". Bien sûr, le fait que les lignes du graphe ne portent pas seulement des masses mais des noms de particules est une distinction un peu byzantine Une différence plus importante est illustrée par la figure ci-dessous

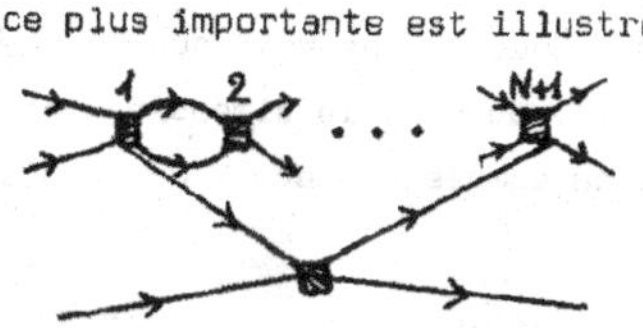

(N = 1,2,...)

qui représente une infinité de graphes tous associés au même type de configuration causale. Ce phénomène, qui se retrouve évidemment chaque fois qu'un graphe a des lignes multiples, n'a aucune importance pour les considérations du présent paragraphe*), mais conduira à des difficultés techniques (que nous éluderons) dans les énoncés du paragraphe suivant.

2.1. Région physique et variété de Landau d'un graphe de diffusion multiple.

Soit G un graphe de diffusion multiple, dont l'ensemble de lignes (internes et externes) sera noté $|G|$. Formons le produit $\prod_{i\in|G|} M_i$ de toutes les couches de masse des particules en jeu, et coupons le par les plans (de codimension 4) définis par la conservation de l'impulsion-énergie à chaque sommet de G : on obtient ainsi une variété algébrique ("généralement" lisse) que nous noterons M_G, et que nous appellerons couche de masse du graphe G. Si I resp. J désigne l'ensemble des lignes externes entrantes resp. sortantes du graphe G, on a une projection évidente $\pi_G : M_G \longrightarrow M_{(IJ)}$ définie en "oubliant" les impulsions-énergies des lignes internes. Il est facile de voir que π_G est un morphisme propre (car en bornant les énergies des particules externes du graphe on borne les énergies des particules internes). L'image de ce morphisme est donc un fermé semi-algébrique de $M_{(IJ)}$, qui sera appelé région physique du graphe G. Nous noterons Γ_G l'ensemble critique de π_G (ensemble des points où l'application tangente n'est pas surjective) et $L(G)$ le "contour apparent" de π_G, c.à.d. l'image de Γ_G par π_G. Ce "contour apparent" $L(G)$ est un fermé semi-algébrique (nulle part dense) de $M_{(IJ)}$, que nous appellerons "variété de Landau" du graphe G, et dont les points seront appelés "points de Landau".

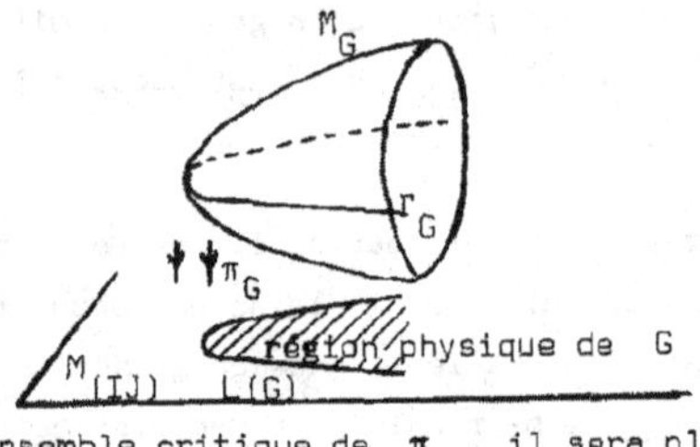

Pour calculer l'ensemble critique de π_G, il sera plus commode de raisonner en termes d'application cotangente : nous allons voir que le noyau de l'application cotangente en un point critique (noyau dont la dimension est le "corang" du point critique) admet une interprétation simple en termes de "configurations" dans l'espace-temps.

---

*) Les objets définis ci-après ne dépendent en fait que du type de configuration causale associée à G.

2.2. Configurations critiques et covecteurs critiques d'un graphe de diffusion multiple.

On appellera configuration libre toute famille de droites de l'espace-temps, de genre temps et orientées dans le sens du temps. En particulier à tout point $(p) \in M_{(IJ)}$ est associée une "configuration libre élémentaire", déduite de la configuration 1.2 en prolongeant les demi-droites par des droites. D'après la description 1.1 de l'espace cotangent, on peut identifier tout covecteur $(u) \in T^*_{(p)}M_{(IJ)}$ à une translation relative des lignes de cette configuration libre, qui la transforme en une autre configuration libre (dite "associée à(u)").

On appellera configuration critique d'un graphe de diffusion multiple G toute configuration libre, dont les lignes sont indexées par $|G|$, et qui à l'étoile de chaque sommet de G fait correspondre une configuration libre élémentaire (c.à.d. des droites concourantes satisfaisant à la loi de conservation de l'impulsion-énergie).

Un covecteur $(u) \in T^*_{(p)}M_{(IJ)}$ sera appelé covecteur critique de G si la configuration libre qui lui est associée est la famille des lignes externes d'une configuration critique de G.

PROPOSITION : L'ensemble des covecteurs critiques de G n'est autre que le noyau de l'application cotangente à π_G.

La démonstration de cette proposition est immédiate une fois qu'on a compris la description suivante de l'espace cotangent à M_G en un point $(p) \in M_G$:

Pour chaque "étoile" e (processus élémentaire) du graphe G on considère le point $(p)_e \in M_e$, image de (p) par la projection évidente, et l'on associe à ce point une configuration libre élémentaire L_e (de la façon déjà indiquée); soit $L = \coprod_{e \in \text{ét } G} L_e$ la configuration libre définie par l'union disjointe (pour toutes les étoiles de G) de ces configurations élémentaires ; il faut bien noter que chaque ligne interne du graphe G apparaît deux fois dans L, une fois pour l'étoile d'où elle sort et une fois pour l'étoile où elle entre : ce qui fait qu'à chaque ligne interne de G est associé dans L un couple de droites parallèles. L'espace cotangent à M_G en (p) peut s'identifier à l'espace des translations des lignes d'une telle configuration L, modulo translation d'ensemble de chaque L_e et modulo translation d'ensemble de chacun des couples de droites parallèles dont nous venons de parler.

2.3 . Configurations causales "associées" et "subordonnées" à un graphe.

Parmi les configurations critiques que nous venons d'introduire, certaines sont "acausales" : nous entendons par là que leurs sommets sont disposés dans l'espace-temps suivant un ordre temporel non conforme à l'orientation du graphe G . Les autres, tout en respectant l'ordre temporel, peuvent éventuellement être "dégénérées" en ce sens que plusieurs sommets de G peuvent s'envoyer sur le même point de $\mathbb{R}^4$; nous les appellerons configurations causales subordonnées au graphe G ; parmi elles, celles qui ne dégénèrent pas seront appelées configurations causales associées à G : ce sont les configurations libres déduites (par prolongement des lignes) de configurations causales du type 1.3. De façon générale, à toute configuration causale subordonnée à G correspond une "contraction" G' de G (si deux sommets de G ont même image dans $\mathbb{R}^4$, on identifie ces deux sommets et l'on contracte les lignes qui les joignaient) telle que la configuration causale puisse être considérée comme associée à G' (en effaçant les droites qui correspondaient aux lignes contractées). On voit ainsi que tout covecteur causal subordonné à G est un covecteur causal associé à une contraction G' de G .

2.4. Structure de la singularité de Landau au voisinage d'un point causal de corang 1

Soit $p_c \in M_G$ un point critique causal de corang 1 associé au graphe G : on entend par là un point critique de corang 1 de π_G (i.e. Ker $T^*_{p_c}\pi_G$ est un espace vectoriel à une dimension) tel que la configuration critique correspondante (unique à dilatation près) soit une configuration causale associée à G . Soit $p_L \in M_{(IJ)}$ le point de Landau, image de p_c . Dans ces conditions on peut démontrer la

PROPOSITION :

1°) La fibre de π_G au-dessus de p_L se réduit au seul point p_c . Au voisinage de p_L , la variété de Landau L(G) est une hypersurface lisse, image isomorphe par π_G de l'ensemble critique Γ_G . Cette hypersurface lisse borde la région physique de G , c.à.d. que l'image de π_G est située d'un seul côté de cette hypersurface ; plus précisément, si l'on choisit le signe d'une équation locale s de L(G) de telle façon que le covecteur ds soit causal , l'image de π_G se trouve du côté $s \geq 0$.

2°) Au voisinage de p_c , l'application π_G est du type "pli" de Thom , c.à.d. qu'on peut choisir sur M_G des coordonnées analytiques locales $x_1,\dots,x_n$ telles que l'application π_G s'écrive dans ces coordonnées

$$\begin{aligned} y_1 &= x_1 \\ &\dots \\ y_k &= x_k \\ s &= x_{k+1}^2 + \dots + x_n^2 \quad (\text{ou} = 0 \text{ dans le cas } n = k) \end{aligned}$$

[$y_1, \dots, y_k$ sont des coordonnées locales sur $L(G)$] .

2.5. La structure des points critiques de corang > 1 est beaucoup moins bien connue. Cependant la prééminence du corang 1 est attestée par le résultat suivant :

PROPOSITION : Si G est un graphe connexe, l'ensemble des covecteurs causaux subordonnés à G en un point $p_L \in L(G)$ est un cône convexe à base polyédrale, dont les points extrêmaux (sommets du polyèdre) sont des covecteurs causaux de corang 1 associés à l'une des contractions de G .

NOTATION .

$L^+(G)$ = partie causale de $L(G)$ = ensemble des points de Landau d'où part au moins un covecteur causal subordonné à G ($L^+(G)$ est un sous-fermé semi-algébrique de $L(G)$) .

$L_1^+(G)$ = partie causale de corang 1 de $L(G)$ = ensemble des points de Landau du type 2.4 .

REMARQUE . $L_1^+(G)$ peut être vide : si le graphe G a beaucoup de lignes internes et peu de cycles, on peut avoir $\dim M_G < \dim M_{(IJ)} - 1$, et alors aucun point n'est de corang 1 .

Cependant la proposition précédente montre que pour tout graphe G connexe

$$L^+(G) \subset \bigcup_{G'} L_1^+(G')$$, où l'union porte sur toutes les contractions G' de G .

2.6. Singularités génériques des amplitudes de diffusion.

Si nous appelons point causal de $M_{(IJ)}$ tout point d'où part un covecteur causal, l'hypothèse de microanalyticité du §1 nous dit qu'en un point non causal l'amplitude de diffusion $S_{(IJ)}$ est microanalytique dans toutes les directions, donc analytique :

(a) $S_{(IJ)}$ est analytique dans $M_{(IJ)} - L^+_{(IJ)}$, où $L^+_{(IJ)}$ désigne la partie causale de $M_{(IJ)}$.

Or les considérations qui précèdent nous montrent que $L^+_{(IJ)} = \bigcup_G L^+(G)$,
où l'union peut être considérée comme localement finie grâce à 1.5 .
Il en résulte que $L^+_{(IJ)}$ est un fermé, localement semi-algébrique, nulle part dense dans $M_{(IJ)}$.

Demandons nous maintenant quel type de singularité peut avoir $S_{(IJ)}$ en un point générique de $L^+_{(IJ)}$. Il résulte de 2.4 et 2.5 que si l'on oublie la partie de $L^+_{(IJ)}$ qui provient des graphes non connexes ("oubli" qui trouvera sa justification dans l'Appendice) , $L^+_{(IJ)}$ est au voisinage de presque tout point une hypersurface lisse, d'équation locale s , telle qu'en tout point de cette hypersurface ds soit l'unique covecteur causal*) . L'hypothèse de microanalyticité se traduit alors de la façon suivante :

(b) au voisinage d'un tel point "générique" de $L^+_{(IJ)}$, l'amplitude de diffusion $S_{(IJ)}$ est valeur au bord d'une fonction f_+ analytique "du côté" $\mathrm{Im}\, ds > 0$.

Joint à (a) , ce résultat (b) montre en particulier que les deux fonctions analytiques que définit $S_{(IJ)}$ dans les deux demi-espaces réels $U \cap \{s < 0\}$ et $U \cap \{s > 0\}$ (U = voisinage d'un point causal générique dans $M_{(IJ)}$) se déduisent l'une de l'autre par prolongement analytique dans un "tube local" $U \times \Omega^+$ dont la base imaginaire Ω^+ est un ouvert du demi-espace $\mathrm{Im}\, s > 0$. "tangent" à l'hyperplan $\mathrm{Im}\, s = 0$

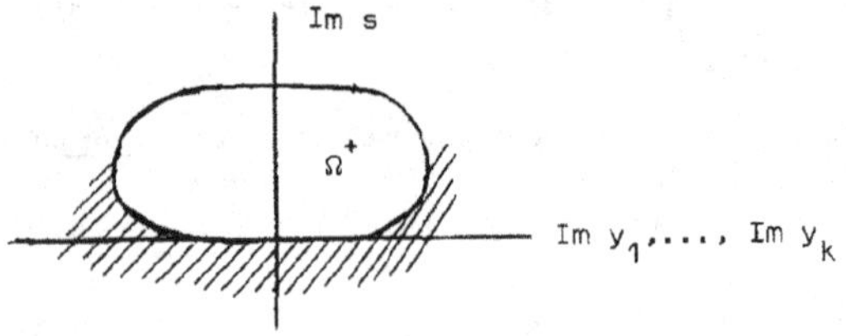

3. SPECTRE DES AMPLITUDES DE DIFFUSION.

L'hypothèse de microanalyticité ne nous donnait de renseignements que sur la position des singularités de l'amplitude de diffusion (dans le fibré cotangent). "L'hypothèse spectrale" qui va suivre nous renseignera sur la nature de ces singularités.

*) Si deux graphes G_1 et G_2 ont (localement) même variété de Landau $L^+ = L^+(G_1) = L^+(G_2)$, il est facile de voir que leurs régions physiques sont du même côté de L^+ , de sorte que si ds est covecteur causal pour G_1 c'est encore ds (et non $-ds$) qui est causal pour G_2 .

DEFINITIONS-NOTATIONS

3.1 On appelle élément de matrice S d'un graphe de diffusion multiple G , et l'on note $S_{<G>}$, la distribution sur la variété $\prod_{i \in |G|} M_i$ (produit des couches de masse de toutes les particules internes et externes) définie, dans le cas où G n'a pas de lignes multiples, comme le produit des éléments de matrice S de tous les processus élémentaires dont est constitué le graphe G (étoiles des sommets de G) . Si G a des lignes multiples il y a lieu de modifier légèrement cette définition, d'une façon que nous ne préciserons pas ici.

3.2 Soit I (resp. J) l'ensemble des lignes externes entrantes (resp. sortantes) du graphe G . On note $A_{IJ}(G)$ la distribution sur $\prod_{i \in I \sqcup J} M_i$ définie en intégrant $S_{<G>}$ sur le produit des couches de masses des particules internes de G (chaque couche de masse M_i est munie de la mesure invariante

$$\delta(p^2 - m_i^2)\, dp_{(o)} dp_{(1)} dp_{(2)} dp_{(3)})\quad .$$

3.3 Il est clair que les distributions définies ci-dessus peuvent être divisées par une fonction δ de Dirac exprimant la conservation de l'impulsion-énergie. On écrira

$$S_{<G>} = \delta(\sum_{i \in I} p_i - \sum_{j \in J} p_j)\, S_G$$

$$A_{IJ}(G) = (\sum_{i \in I} p_i - \sum_{j \in J} p_j)\, A_{(IJ)}(G)$$

où S_G (resp. $A_{(IJ)}(G)$) est une distribution définie sur la couche de masse M_G (resp. $M_{(IJ)}$) .

Pour abréger on notera

$$A(G) = A_{(IJ)}(G)$$

qu'on appellera partie absorptive de l'amplitude de diffusion $S_{(IJ)}$, relativement au graphe G .

Remarquons que $A(G)$ s'obtient en intégrant S_G le long des fibres de la projection $\pi_G : M_G \longrightarrow M_{(IJ)}$ (avec une mesure invariante facile à écrire). En particulier $A(G)$ a son support dans la région physique du graphe G .

3.4 REMARQUE : Les définitions ci-dessus comportent des opérations a priori "dangereuses" s'agissant de distributions : produits, intégrales le long des fibres...

Heureusement, grâce à ce que nous savons du "support spectral" de ces distributions (hypothèse de microanalyticité) des arguments standard de théorie des hyperfonctions (cf. exposés de Chazarain) permettent de montrer facilement que toutes les opérations ci-dessus ont un sens, et de préciser le "support spectral" des distributions ainsi obtenues. On obtient notamment le résultat suivant

PROPOSITION : La partie absorptive $A(G)$ est microanalytique partout, sauf dans les codirections associées à des configurations obtenues en "éclatant de façon causale" les sommets de G : on entend par là les configurations critiques $\hat{C}$ associées à des graphes $\hat{G}$ tels que

1°) G soit donné comme contraction de $\hat{G}$;

2°) chaque sous-graphe de $\hat{G}$ qui se contracte suivant une étoile (processus élémentaire) de G est associé dans $\hat{C}$ à une sous-configuration causale.

En particulier, si $(u) \in T^*M_{(IJ)}$ est un covecteur causal simple (au sens de 1.5) associé au graphe de diffusion multiple G, il résulte de la proposition ci-dessus que le support spectral de $A(G)$ coïncide au voisinage de (u) avec l'ensemble des covecteurs critiques de G (tous causaux puisque voisins de (u)). Autrement dit, les supports spectraux de $A(C)$ et de $S_{(IJ)}$ coïncident au voisinage de (u). "L'hypothèse spectrale" ci-dessous dit que non seulement les supports spectraux mais aussi les "spectres" (au sens de SATO, c.à.d. les microfonctions correspondantes) sont les mêmes.

HYPOTHESE SPECTRALE :

Soit $(u) \in T^*M_{(IJ)}$ un covecteur causal simple (au sens 1.5) associé à un graphe de diffusion multiple G et un seul. Alors la distribution $S_{(IJ)} - A(G)$ est microanalytique dans la direction du covecteur (u).

3.5 REMARQUE : En réalité, l'hypothèse spectrale n'est pas indépendante de l'hypothèse de microanalyticité, mais peut en être déduite à l'aide de la propriété fondamentale d'unitarité de la matrice S. (cf. exposé d'OLIVE). Si nous avons préféré présenter ces deux "hypothèses" sur un pied d'égalité, c'est parce que toutes deux sont également fondamentales du point de vue de l'interprétation physique (voir l'exposé de Iagolnitzer).

3.6 FORMULES DE DISCONTINUITE.

L'hypothèse spectrale va nous permettre de préciser les informations données au n° 2.6 sur les singularités de l'amplitude de diffusion $S_{(IJ)}$ au voisinage d'un point de Landau $p_L \in L_1^+(G)$, moyennant l'hypothèse supplémentaire que

p_L n'est point de Landau d'aucun autre graphe que G .

Sous cette hypothèse, non seulement l'amplitude de diffusion $S_{(IJ)}$ n'est singulière que dans la direction du covecteur causal (u) , mais la partie absorptive A(G) n'est singulière (Proposition 3.4) que dans les deux co-directions (u) et (-u) (codirections associées aux seules configurations critiques de G : la "causale" (u) et "l'anticausale" (-u)). Par conséquent $S_{(IJ)} - A(G)$ ne peut a priori être singulière que dans ces deux codirections, donc en réalité seulement dans la codirection (-u) puisque l'autre est exclue par l'hypothèse spectrale.

D'où le résultat (avec les notations 2.6) :

(c) au voisinage de p_L , la distribution $S_{(IJ)} - A(G)$ est valeur au bord d'une fonction f_- analytique "du côté" $\mathrm{Im}\, ds < 0$.

Or on sait d'après 3.3 que A(G) s'annule "en dessous du seuil" de G , c.à.d. du côté $s < 0$, complémentaire de la région physique de G . Les deux fonctions analytiques f_+ (définie en 2.6 (b)) et f_- (définie par (c) ci-dessus) se prolongent donc en une même fonction analytique f dont on peut se représenter le domaine d'analyticité comme sur la figure suivante :

plan des Im p , pour Re p en dessous du seuil (c.à.d. Re s < 0).

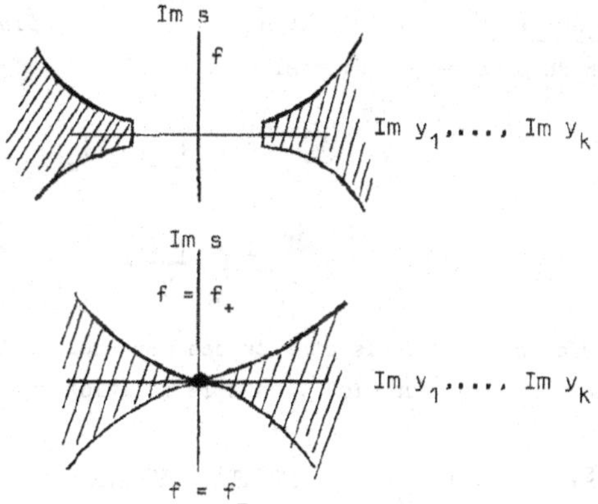

Plan des Im p , pour Re p en dessus du seuil (c.à.d. Re $s \geq 0$).

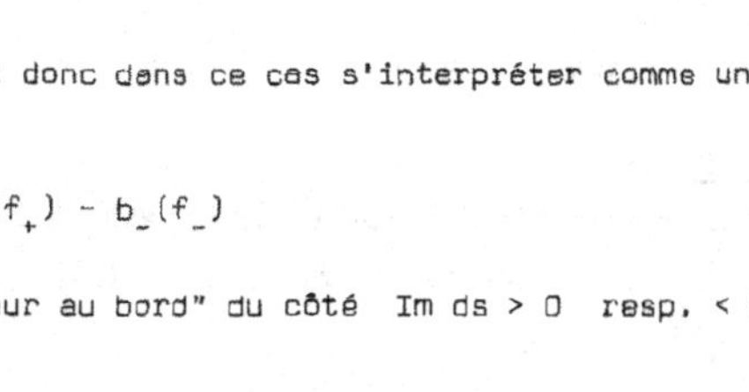

La partie absorptive A(G) peut donc dans ce cas s'interpréter comme une "discontinuité"

$$A(G) = b_+(f_+) - b_-(f_-)$$

(où b_+ resp. b_- signifie "valeur au bord" du côté $\mathrm{Im}\, ds > 0$ resp. < 0) .

APPENDICE : DECOMPOSITION EN PARTIES CONNEXES

Les configurations causales définies en 1.3 ne sont pas nécessairement connexes. Parmi les configurations causales non connexes, les plus simples sont celles qui n'ont pas de lignes internes : ce sont tout simplement des unions disjointes de configurations élémentaires.

Soit $(u) \in T^{*}_{(p)}M_{(IJ)}$ un covecteur associé à une telle configuration

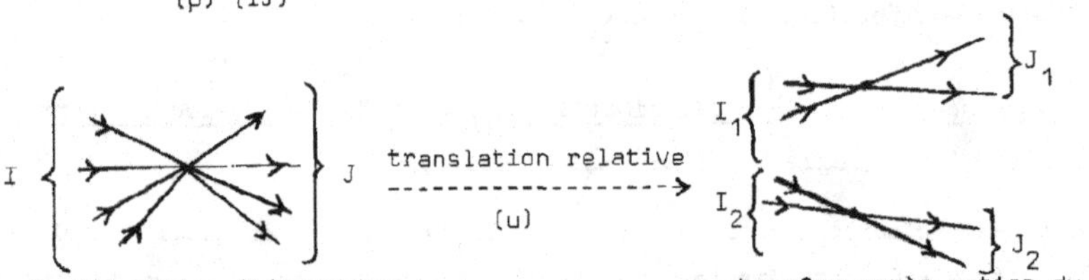

Configuration élémentaire associée à (p).

La même après action du covecteur (u).

Un tel covecteur est associé à un graphe E que nous appellerons "décomposition élémentaire" de (IJ), et qui est tout simplement une union disjointe de graphes de processus élémentaires :

$$E = \coprod_e (I_e \, J_e) \quad ,$$

$$I = \coprod_e I_e \quad , \quad J = \coprod_e J_e \quad .$$

L'hypothèse spectrale dans ce cas particulier très simple dit qu'au voisinage d'un tel covecteur (u), on a la relation "microfonctionnelle" :

(*) $\quad S_{IJ} - \prod_e S_{I_e J_e}$ est microanalytique .

Cette relation donne tout son sel au lemme (purement combinatoire) suivant :

LEMME :

Supposons donnée une famille d'objets $(S_{IJ})_{(IJ)\in\mathcal{E}}$ indexée par l'ensemble $\mathcal{E}$ de tous les processus élémentaires (IJ). A une telle famille est associée de façon unique une famille $(S^c_{IJ})_{(IJ)\in \mathcal{E}}$, où les S^c_{IJ} sont des éléments de l'algèbre libre engendrée par les S_{IJ} , telle que pour tout $(IJ)\in\mathcal{E}$ on ait

$$(**)\qquad S_{IJ} = \sum_{E\in \text{déc}(IJ)} \prod_{e\in \text{ét}(E)} S^c_{I_e J_e}$$

où déc(IJ) désigne l'ensemble des décompositions élémentaires du processus (IJ) , et ét(E) désigne l'ensemble des étoiles (c.à.d. des parties connexes) d'une telle décomposition E .

PREUVE DU LEMME :

En isolant dans le membre de droite de (**) le terme correspondant à la "décomposition triviale" E = (IJ) , on trouve une formule du type

$$S_{IJ} = S^c_{IJ} + \ldots$$

où les ... représentent des sommes de produits de termes $S^c_{I_e J_e}$, où I_e et J_e sont des sous-ensembles stricts de I et J ; le lemme s'en déduit immédiatement par récurrence sur le nombre d'éléments des ensembles I et J .

DEFINITION: Les S^c_{IJ} s'appellent "parties connexes" des S_{IJ} , et la formule (**) s'appelle "formule de décomposition en parties connexes".

Si l'on veut étendre ce lemme (purement combinatoire dans l'énoncé ci-dessus) à notre situation où les S_{IJ} sont des distributions, le problème se pose de savoir si la multiplication de ces distributions a un sens. La réponse est encore une fois donnée, comme au n° 3.4 , par l'hypothèse de microanalyticité, qui nous permet non seulement de donner un sens au produit des distributions mais même de calculer le support spectral des nouvelles distributions ainsi obtenues; on voit ainsi sans difficulté que $\text{Supp Spec } S^c_{IJ} \subset \text{Supp Spec } S_{IJ}$.

Mais il y a mieux : l'hypothèse spectrale et notamment la relation (*) permet de voir que l'inclusion ci-dessus est stricte, le support spectral de S^c_{IJ} n'étant constitué que de covecteurs associés à des configurations causales connexes. De façon précise, on peut démontrer la

PROPOSITION :

Il y a équivalence entre

1°) La donnée d'une famille de distributions $S_{IJ} = \delta(\sum_{i\in I} p_i - \sum_{j\in J} p_j) S_{(IJ)}$, $(IJ)\in\mathcal{E}$, satisfaisant à l'hypothèse de microanalyticité (1.4) ainsi qu'à l'hypothèse spectrale (3.4) , et

2°) la donnée d'une famille de distributions

$$S^{c}_{IJ} = \delta(\sum_{i\in I} p_i - \sum_{j\in J} p_j)\ S^{c}_{(IJ)} \quad , (IJ) \in \mathcal{E} \quad ,$$

satisfaisant à deux hypothèses analogues mais où les "configurations causales" sont partout remplacées par les "configurations causales connexes".

Ces deux familles de donnée sont reliées entre elles par la formule de "décomposition en parties connexes" (**) , qui est bien définie "au sens des distributions".

NOTE BIBLIOGRAPHIQUE

Les axiomes de la matrice S exposés ici sont le résultat d'une lente maturation dont on peut voir l'aboutissement dans l'article de

. D. IAGOLNITZER et H.P.STAPP - Macroscopic causality and physical region analyticity in S-matrix theory, Comm. Math. Phys. 14,15 (1969).

On en trouvera un excellent exposé dans le livre de

. D. IAGOLNITZER - Introduction to S. Matrix Theory.
(Association pour la diffusion de textes scientifiques et littéraires, Paris 1973).

Je n'ai fait ici que reprendre les mêmes idées en remarquant que le langage "microfonctionnel" de SATO permet de leur donner une forme particulièrement concise : pour apprendre à parler ce langage, lire

. Introduction aux hyperfonctions par A.CEREZO, A.PIRIOU, J.CHAZARAIN (dans ce volume)

Voici quelques références complémentaires (avec en regard le § de mon exposé auxquelles elles se rapportent).

(Introduction : "principe de superposition") - R.P. FEYNMAN, R.B. LEIGHTON, M. SANDS The Feynman lectures in Physics, vol I, chap. 37 (Addison-Wesley 1969).

(§1.5) H.P. STAPP - Finiteness of the number of positive α Landau surfaces in bounded portions of the physical region - J. Math. Phys. 8,8 (1967) .

(§2.4) F. PHAM - Singularités des processus de diffusion multiple Ann. Inst. Henri Poincaré 6, 2 (1967) .

(§3.5) D. OLIVE (dans ce volume).

MACROCAUSALITY, PHYSICAL-REGION ANALYTICITY AND INDEPENDENCE PROPERTY IN S-MATRIX THEORY

D. IAGOLNITZER

DPh-T CEN Saclay BP n°2 91190 Gif-sur-Yvette
FRANCE

ABSTRACT

The equivalence between macrocausality and physical-region analyticity properties of the S-matrix, which has first been proved in reference 1, is described here in an improved version derived from Chapter II of reference 2.

This version follows from various mathematical developments (references 3,4) which have allowed to give a somewhat better statement of the macrocausality property and to complete the results of reference 1 in various ways.

It turns out, in particular, that the independence property, which was originally presented (in some situations) as a supplementary assumption, can always be derived from macrocausality.

INTRODUCTION

As mentioned in the previous lecture by Pham[6], the basic quantities in the quantum relativistic physics of systems of massive particles ($m > 0$) with short-range interactions, such as the "strong" interactions, are the scattering amplitudes S_{IJ} between sets I of initial particles and sets J of finalparticles.

We first briefly recall, for the non specialized reader, the physical meaning of S_{IJ} . (For details, see for instance[2]). Since the spin variables are unessential in the topics discussed, we shall only consider spinless particles.

A "pure" (i.e. "completely determined") state of a free particle with mass m is represented in quantum relativistic physics by a "wave function" φ , up to multiplication by a complex constant ; φ is a vector of $L^2(M)$, i.e. is a function of the (real) energy-momentum variable $p \equiv p_0 , \vec{p}$, $p \in M$ ($p_0 > 0$,

$p^2 = p_o^2 - \vec{p}^{\,2} = m^2$), and is square integrable with respect to the measure $d\mu(p) = \delta(p^2-m^2)\,\theta(p_o)\,d^4p \equiv \dfrac{d\vec{p}}{2\sqrt{\vec{p}^{\,2}+m^2}}$:

$$\|\varphi\|^2 = \int |\varphi(p)|^2 \, d\mu(p) < \infty \,. \tag{1}$$

Now, before and after the scattering processes take place, the physical systems under consideration can be asymptotically identified with <u>free particle states</u>. The basic principles of quantum theory then entail the existence, for any given sets I,J, of the corresponding "scattering amplitude" S_{IJ} : S_{IJ} is a functional which acts on the sets $\{\psi_k\} = (\{\psi_i\}_{i\in I}, \{\psi_j\}_{j\in J})$, $\psi_k \in L^2(M_k)$ $(\forall k)$, is <u>linear</u> with respect to each variable ψ_k, and is such that the <u>transition probability</u>(*) from a set I of initial particles in the states φ_i, $i\in I$ to a set J of final particles in the states φ_j, $j\in J$, can be written in the form :

$$W(\{\varphi_k\}) = |S_{IJ}(\{\varphi_i^*\}, \{\varphi_j\})|^2 \times (\prod_k \|\varphi_k\|^2)^{-1} \,, \tag{2}$$

where $*$ denotes complex conjugation.

The inequality $W \leq 1$, i.e. :

$$|S_{IJ}(\{\varphi_k^*\})| < \prod_k \|\varphi_k\| \,, \tag{3}$$

ensures in particular that S_{IJ} is a <u>tempered distribution</u> in momentum space. In view of energy-momentum conservation(**) and of Eq.(3), S_{IJ} can be written in the form :

$$S_{IJ} = S_{(IJ)}\, \delta^4\left(\sum_{i\in I} p_i - \sum_{j\in J} p_j\right) \,, \tag{4}$$

where $S_{(IJ)}$ is defined on the manifold $M_{(IJ)}$ of the points $p = \{p_k\}$ such that $p_k \in M_k$, $\forall k$, and $\sum_{i\in I} p_i = \sum_{j\in J} p_j$; $M_{(IJ)}$ is called the <u>physical-region</u> of the scattering process $I \to J$.

(*) i.e. the probability of detecting a set J of final particles in the states φ_j, starting with initial particles in the states φ_i.

(**) i.e. $W(\{\varphi_k\}) = 0$ if the relation $\sum_{i\in I} p_i = \sum_{j\in J} p_j$ cannot be satisfied with each p_k in the support of φ_k.

Finally it turns out to be useful to introduce the "connected parts" S^c_{IJ} of the functionals S_{IJ} (*) ; they can also be written in the form :

$$S^c_{IJ} = T_{IJ}\, \delta^4\left(\sum_{i\in I} p_i - \sum_{j\in J} p_j\right) \tag{4'}$$

where T_{IJ} is defined, as $S_{(IJ)}$, on $M_{(IJ)}$.

Macrocausality[7] is an appropriate mathematical expression of a certain classical limit of quantum theory, namely of the principle that any energy-momentum transfer over large distances which cannot be attributed to (stable) physical particles according to classical ideas, gives effects that are damped exponentially with distance. (This includes in particular the idea of the short-range character of the interactions).

To get a better understanding of the classical ideas involved, it is first necessary to abandon quantum theory and to study, as a guide, a classical model of point particles. This model has already been introduced in PHAM's lecture.

We review it in section I, where we indicate the main definitions and results about the $+\alpha$ - Landau surfaces and the causal displacements[8] .

In section II, we return to quantum theory, and indicate how the classical ideas can be adapted, asymptotically, in the quantum case. The macrocausality property is then stated in the form of exponential fall-off properties of the transition probabilities, under appropriate conditions, when subgroups of initial and final particles are displaced from each other.

We show that this property amounts to the following basic result : the "essential support" of $S_{(IJ)}$, resp. of T_{IJ} , at a point $p = \{p_k\}$ of $M_{(IJ)}$ is (contained in) the set of causal directions at p as determined in the classical model of section I, resp. is (contained in) the set of causal directions at p associated with connected configurations.

(*) The connected parts are defined, by induction, by the formulae :

$$S_{IJ} = S^c_{IJ} + \sum_{\pi_1 \dots \pi_\ell} \bigotimes_{t=1}^{\ell} S^c_{I_t J_t} ,$$

where the sum Σ runs over all non trivial partitions of (I,J) into subsets (I_t, J_t) and $\otimes$ is essentially the tensorial product.

The definition of the essential support of a distribution is recalled in Appendix I,[11] where the main results needed below are described. (For more details, see [3]).

We also see there that the essential support coincides with the "singular support", also called "singular spectrum" , introduced independently, by very different methods, by Professor SATO and coworkers[5], and that the micro-analyticity property indicated by PHAM therefore coincides with the above essential support property of $S_{(IJ)}$, resp. T_{IJ} .

In section III, we finally describe the analyticity properties implied by, and as a matter of fact equivalent to, the above essential support property :

a) For any given I,J, there exists a unique analytic function F_{IJ} (defined in a domain of the complexified manifold $\overline{M}_{(IJ)}$ of $M_{(IJ)}$) to which T_{IJ} is equal at all non $+\alpha$ Landau points, and from which it is a "plus $i\varepsilon$" boundary value at almost all $+\alpha$ -Landau points.

Besides a few exceptional points, the remaining $+\alpha$ - Landau points are (some of) those which lie in the intersection of several $+\alpha$-Landau surfaces with no "common parent". For them, the following property was still derived in [1] :

b) In the neighborhood of a $+\alpha$-Landau point p of the latter type, T_{IJ} is a <u>sum</u> of boundary values (in the sense of <u>distributions</u>) of analytic functions G_β . Each of these boundary values is obtained from "plus $i\varepsilon$" directions associated with one of the "parent" surfaces involved at p .

The "independence property in its original form was the assertion "that each G_β has moreover, <u>in the neighborhood of p</u> , the same analytic properties as if no other parent surface was involved, "(i.e. G_β is for instance analytic at all real points p' which do not lie on the parent surface considered, or possibly on its own "daughters"). This property was proved in [1] only in special situations, this being due to the lack of general information on the links between the various decompositions of a distribution, obtained at different points p , into sums of boundary values of analytic functions.

More recent mathematical results[4] provide this information and the independence property easily follows. As a matter of fact, these results also provide global decomposition properties of T_{IJ} in $M_{(IJ)}$.

I - CAUSALITY IN A CLASSICAL MODEL [8] : + α - LANDAU SURFACES AND CAUSAL DISPLACEMENTS

The classical model considered obeys the following laws :

i) A pure state of a free particle is characterized by one energy-momentum 4-vector p, $p \in M$ ($p_o > 0$, $p^2 = m^2$) , and by the position $\vec{x}_o$ of the particle at one time t_o . The particle has a well defined space-time trajectory which is the line parallel to p and passing through the point $(t_o, \vec{x}_o)$, according to the propagation law :

$$\frac{\Delta \vec{x}}{\Delta t} \equiv \vec{v} = \frac{\vec{p}}{p_o} ,$$

where $\vec{v}$ is the velocity.

ii) Particles may "interact" when their space-time trajectories meet at some point. Then, new outgoing particles, which replace the incoming ones, may emerge from the interaction point with the same properties as before, until their space-time trajectory meets again that of another particle, and so forth.

At each interaction point, energy-momentum must be conserved (i.e. the sum of the incoming energy-momentum 4-vectors must equal that of the outgoing ones), and the number of outgoing (as well as incoming) particles must be larger than or equal to 2.

Being given sets I and J of initial and final particles and a corresponding set $p \equiv \{p_k\}$ of 4-momenta, we then consider sets $\mathcal{U} \equiv \{u_k\}$ of space-time displacements : starting from trajectories which all pass through the origin, the displaced trajectory of particle k now passes through u_k .

$\mathcal{U}$ is said to be causal at p , or the set $(p, \mathcal{U})$ is said to be causal, if energy-momentum can be transferred from the initial to the final particles, possibly by intermediate particles, according to the laws i) and ii) above. A causal $\mathcal{U}$ is moreover said to be connected if it corresponds to a connected configuration of all (initial, final and intermediate) particles involved.

The remainder of this section is devoted to the study of the connected causal sets.

+α - Landau surfaces

If $(p, \mathcal{U})$ is causal, p must clearly belong to the submanifold $M_{(IJ)}$ of $\mathbb{R}^{4(|I|+|J|)}$ defined by the conditions

$$p_k \in M_k : p_k^2 = m_k^2 \ , (p_k)_o > 0 \ , \forall k$$

$$\sum_{i \in I} p_i = \sum_{j \in J} p_j \ . \tag{4}$$

$M_{(IJ)}$ is a smooth manifold except at the exceptional points (when they exist) where all initial and final 4-momenta are colinear. These points will be excluded below.

To study the connected causal sets $(p, \mathcal{U})$, one first considers all <u>connected topological graphs</u> G with oriented external and internal lines such that :

i) Their incoming, resp. outgoing, external lines are in a 1-1 correspondence with the elements $i \in I$, resp. $j \in J$, and, for each point $p \equiv \{p_k\}$ considered, with the corresponding 4-momenta p_k.

ii) To each internal line ℓ, is attributed the mass m_ℓ of a physical particle.

Finally, there are at least 2 incoming and 2 outgoing lines at each vertex.

A point $p \equiv \{p_k\}$ is said to belong to the <u>$+\alpha$ Landau surface</u> $L^+(G)$, resp. $L_o^+(G)$, if it is possible to find a set of 4-momenta k_ℓ for each internal line ℓ, $k_\ell \in M_\ell$ $(k_\ell^2 = m_\ell^2, (k_\ell)_o > 0)$ and a set of $\alpha_\ell \geq 0$, resp. $\alpha_\ell > 0$, such that :

a) energy-momentum is conserved at each vertex of G

b) $\sum_\ell z(\ell)\, \alpha_\ell\, k_\ell = 0$ for each closed loop z of G : $z(\ell) = 0$ if z does not contain line ℓ, $z(\ell) = +1$, resp. -1, if it contains it with the correct, resp. opposite, orientation. Eqs (a)(b) are called the $+\alpha$-Landau equations of G.

It can be checked that the surfaces $L_o^+(G)$ (if not empty) are <u>analytic submanifolds of codimension 1</u>(*) of $M_{(IJ)}$, and that their union is <u>not</u> dense in $M_{(IJ)}$, but divide it into sectors.

Each surface $L_o^+(G)$ has also a well defined "physical side" in $M_{(IJ)}$ from which it is the boundary, as described in PHAM's lecture(**) :

(*) i.e. their dimension is that of $M_{(IJ)}$ minus one.

(**) The "physical side" lies in the "physical region of G" introduced by PHAM when the latter does not reduce to $L_o^+(G)$ itself. This last situation occurs for graphs G without closed loops, in which case the "physical side" of $L_o^+(G)$ is easily defined directly .

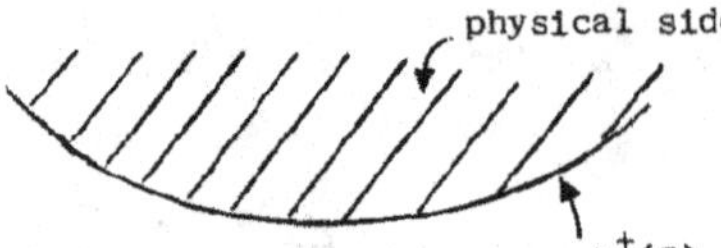

Figure 1

Connected causal displacements

Being given a point $p \equiv \{p_k\}$, the displacements of the form $T = \{\lambda_k p_k + a\}$ are called trivial at p: clearly if $\mathcal{U} = \{u_k\}$ is causal at p, $\mathcal{U}+T$ is also causal since the displacement $\lambda_k p_k$ does not change the trajectory of particle k, and a is a global translation of all trajectories. It will therefore be useful to introduce, at each point $p = \{p_k\}$, the vector space of displacements $\tilde{\mathcal{U}}$ defined only up to addition of trivial displacements at p.

As a matter of fact, the trivial sets T at p are the sets $\mathcal{U} = \{u_k\}$ which are conormal at p to the manifold $M_{(IJ)}$, if the scalar product $\langle \mathcal{U}, \pi \rangle$ of a vector $\mathcal{U} = \{u_k\}$ with a vector $\pi = \{\pi_k\}$ in energy-momentum space is defined by :

$$\langle \mathcal{U}, \pi \rangle = \sum_{i \in I} u_i . \pi_i - \sum_{j \in J} u_j . \pi_j \tag{5}$$

where $u_k . \pi_k = (u_k)_o (\pi_k)_o - \vec{u}_k . \vec{\pi}_k$ (see figure 2).

As mentioned by PHAM, the above space of displacements $\tilde{\mathcal{U}}$ at p can be correspondingly identified with the cotangent vector space $T^*_p M_{(IJ)}$ at p to $M_{(IJ)}$.

The basic facts about the connected causal sets $\mathcal{U} = \{u_k\}$ at a point $p = \{p_k\}$ of $M_{(IJ)}$ can be stated as follows :

a) if p lies on no $+\alpha$ -Landau surface, then there is no (non zero) connected causal $\tilde{\mathcal{U}}$ at p (i.e. all connected causal $\mathcal{U}$ are trivial).

(It is in fact immediately checked that the $+\alpha$- Landau equations follow from the kinematical equations of the model).

b) If p lies on only one surface $L_o^+(G)$ (of $M_{(IJ)}$), then there is only one connected causal direction $\tilde{\mathcal{U}}$ at p (i.e. one $\tilde{\mathcal{U}}$ up to multiplication by a positive scalar λ). This direction is <u>normal at p</u> to $L_o^+(G)$ (in the sense of (5))(*), and is oriented towards the physical side of $L_o^+(G)$.

This situation is represented (locally) in Figure 2, where $M_{(IJ)}$ is (locally) represented by the plane (x,y) and $M^{\perp}_{(IJ)}(p)$ by the z-axis. ($\mathcal{U}$-space has been identified with the space $\mathbb{R}^{4(|I|+|J|)}$ where $M_{(IJ)}$ is embedded, the axis $(p_k)_\nu$ being identified (according to (5)) with $\varepsilon_k(u_k)_\nu$, $\nu = 0,1,2,3$, where $\varepsilon_k = +1$ if k is initial, $\varepsilon_k = -1$ if k is final).

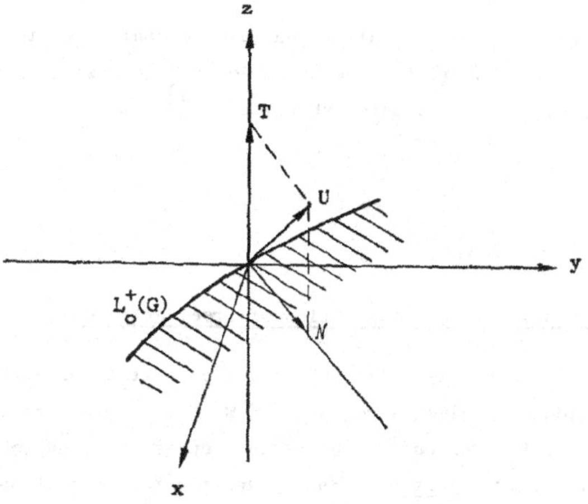

<u>Figure 2</u>

c) If p lies on one surface $L^+(G)$ and on one or more other surfaces $L_o^+(G')$ such that all G' are "contractions" of G(**), then the set of connected causal $\tilde{\mathcal{U}}$ at p is the set of linear combinations, with positive coefficients, of the various $\tilde{\mathcal{U}}$ associated with each $L_o^+(G')$ as in b).

It can be proved that the convex cone of directions obtained is always strictly contained in an open half-space.

(*) I.e. any representative $\mathcal{U}$ of $\tilde{\mathcal{U}}$ is normal at p to $L_o^+(G)$, in $\mathbb{R}^{4(|I|+|J|)}$.

(**) A contraction of G is a graph obtained from G by removing some internal lines and then identifying the end-point vertices of each line removed. This situation occurs when one subset, or various different subsets, of coefficients α_ℓ associated with the lines of G may vanish at p.

d) Finally, if p lies in the intersection of several $+\alpha$ Landau surfaces $L_o^+(G')$, where the graphs G' are not contractions of a common "parent graph" G, then the set of connected causal $\tilde{u}$ is the union of the sets associated with each graph, or parent graph, involved at p.

In some situations, the directions, or convex cones of directions, associated with each graph, or parent graph, involved at p, are disjoint ; p is then called a "type I-point". It is called a "type II-point" otherwise.

In either case, the set of connected causal directions is no longer always contained in a convex (salient) cone. Examples of these situations are easily obtained(*), and some of them have been exhibited in [9].

II - MACROCAUSALITY PROPERTY

Macroscopic space-time localization of free particles [1][2]

In quantum relativistic physics, a pure state of a free particle is no longer represented by a given 4-vector $p \in M$ and a given space-time trajectory parallel to p, but turns out to be instead represented, up to multiplication by a complex constant, by a vector of an irreducible representation space of the covering group of the Poincaré group(**). These representations are in general labelled by two numbers, the mass m, which is strictly positive in the physical cases that we consider, and the spin s which is integer or half-integer.

For spinless particles ($s=0$), the representation space is the space $L^2(M)$ introduced at the beginning of this text, and the representation $\varphi \to \varphi^u$ of the

---

(*) Take for instance a point $p = \{p_k\}$ of $M_{(IJ)}$ such that energy-momentum conservation is satisfied by two subsets of initial and final 4-momenta

$$\left(\sum_{i\in I_1} p_i - \sum_{j\in J_1} p_j = \sum_{i\in I_2} p_i - \sum_{j\in J_2} p_j = 0 \,,\; I_1 \cup I_2 = I \,,\; I_1 \cap I_2 = \emptyset \atop J_1 \cup J_2 = J \,,\; J_1 \cap J_2 = \emptyset \right) .$$

The point p is then in general a "type II-point", whose set of connected causal directions is not contained in a convex salient cone.

(**) The Poincaré group is a semi-direct product of the group of space-time translations and of the group of Lorentz transformations.

space-time translation by a 4-vector u is given by :

$$\varphi^u(p) = \varphi(p)\, e^{ip.u} \tag{6}$$

(where $p.u = p_o u_o - \vec{p}\,\vec{u}$). As a matter of fact, the space-time translations of free particle states have been assumed at the outset to be physically well defined for any 4-vector u, and φ^u is the wave function representing the state obtained from φ after translation by u.

Now, let us first consider the non relativistic quantum case. Then φ is a function of the 3-momentum $\vec{p}$ and the following facts are well known : if φ has a unit norm $\left(\int |\varphi(\vec{p})|^2\, d\vec{p} = 1 \right)$, then $|\varphi(\vec{p})|^2$ is the probability density for detecting the particle with momentum $\vec{p}$ (independently of its position), and $|\tilde{\varphi}(x,t)|^2$ where

$$\tilde{\varphi}(\vec{x},t) = \int \varphi(\vec{p})\; e^{-i(\vec{p}^{\,2}/2m)t}\; e^{i\vec{p}\,\vec{x}}\; d\vec{p} \quad , \tag{7}$$

is the probability density for detecting, at time t, the particle at $\vec{x}$ (independently of its momentum).

In the relativistic quantum case, $|\varphi(p)|^2$ is again the probability density with respect to momentum. On the other hand, the concept of a well-defined space-time position of a particle on the microscopic level is now to be abandoned (for physical and related mathematical reasons). However one may still define a space-time wave function, analogous to $\tilde{\varphi}$, through the formula :

$$f(x) = \int \varphi(p)\, e^{-ip.x}\, d\mu(p) \qquad (x \equiv x_o, \vec{x}) \quad . \tag{8}$$

($d\mu(p) = \delta(p^2-m^2)\,\theta(p_o)\,d^4p$) , or other related quantities and, although these quantities have no longer any interpretation, in general, in terms of probability densities, some features of the non relativistic case do remain valid, on the macroscopic level. We below describe the properties which will be needed later and which are a refined expression of the idea that the probability of finding the particle in macroscopic space-time regions is "negligibly small" if $f(x)$ is itself "negligibly small" in these regions.

Consider a set of wave functions $\varphi_\tau(p)$ of the form :

$$\varphi_\tau(p) = \chi(p)\; e^{-\gamma\tau\,(\vec{p}-\vec{P})^2} \tag{9}$$

where χ is C^∞ (i.e. infinitely differentiable), has a compact support around the point $P = (\sqrt{\vec{P}^2+m^2}\ ,\ \vec{P})$, and is moreover locally analytic at P , and where γ is a positive constant ($\gamma \geq 0$) .

The corresponding space-time wave functions will be denoted by f_τ as above.

Let $V(p)$ be the line issuing from the origin in space-time and parallel to a given 4-vector $p \in M$ ($p_o > 0$, $p^2 = m^2$) and let $V(\chi)$ be the (closed) set of all $V(p)$ with p in the support of χ ; $V(\chi)$ is called the velocity cone of χ :

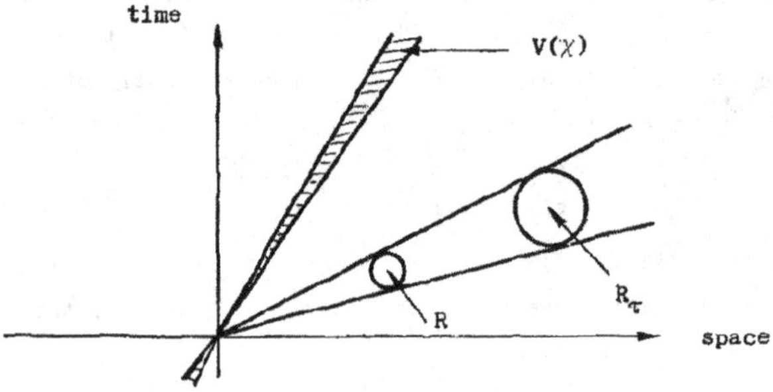

Figure 3

Using the methods of lemma 1a) in Appendix I, one checks[1][2] that, being given any open region R in space-time whose closure does not intersect $V(\chi)$, the following bounds are satisfied for all positive integers N and all sufficiently small γ ($0 \leq \gamma < \gamma_R, \gamma_R > 0$) :

$$\underset{x \in R_\tau}{\text{Max}} \ |f_\tau(x)| < \frac{D_N}{1+\tau^N} \ e^{-\beta\gamma\tau} \ , \ \beta > 0 \ , \tag{10}$$

where R_τ is the region obtained from R by the transformation $x \to \tau x$.

The constant $\beta > 0$ depends only on the real analyticity domain of χ around P . The constants D_N (and γ_R) depend on χ and R, but not on γ or τ.

We note that a similar bound is also obtained at $N = 0$ whenever the closure of R does not intersect the line $V(P)$ itself :

$$\underset{x \in R_\tau}{\text{Max}} \ |f_\tau(x)| < D \ e^{-\beta\gamma\tau} \ ; (0 \leq \gamma < \gamma_R) \tag{10'}$$

where $\beta > 0$ may now depend on R, (if the closure of R intersects $V(\chi)$).

If we next consider the displaced wave functions $\varphi_\tau^{\tau u}$ obtained from φ_τ by the translation τu, where u is a given 4-vector, (see Eq. (6)), it is readily checked (in view of the identity $f_\tau^{\tau u}(\tau u') = f_\tau(\tau(u'-u))$) that $f_\tau^{\tau u}$ again satisfies the bounds (10)(10') whenever the closure of R does not intersect the displaced velocity cone $V^u(\chi)$, whose apex is now the point u instead of the origin, resp. the displaced line $V^u(P)$.

According to the idea on macroscopic space-time localization mentioned earlier, it is correspondingly assumed that the probability $\mathcal{P}(R_\tau)$ of finding the particle whose wave function is $\varphi_\tau^{\tau u}$ in the region R_τ (which is macroscopic for large τ) also possesses analogous exponential fall-off properties in the $\tau \to \infty$ limit.

Macroscopic causality[7]

Consider a set $\{\varphi_{k\tau}^{\tau u_k}\}$ of displaced initial and final wave functions of the above mentioned form :

$$\varphi_{k\tau}^{\tau u_k}(p_k) = \chi_k(p_k)\; e^{-\gamma\tau(\vec{p}_k-\vec{P}_k)^2}\; e^{i\tau(p_k \cdot u_k)} \quad . \tag{11}$$

Then, the initial and final particles have, in the $\tau \to \infty$ limit, properties which are analogous to those of the classical model of section I, with the 4-momentum p_k being possibly replaced by the set of all 4-momenta p_k in the support of χ_k :

i) The probability of finding particle k with a 4-momentum p'_k vanishes if p'_k does not lie in the support of χ_k, resp. decreases exponentially as $e^{-2\gamma\tau(\vec{p}\,'_k-\vec{P}_k)^2}$, in the $\tau \to \infty$ limit, if p'_k lies in that support but is still different from P_k.

ii) The "space-time trajectory" of particle k is the displaced velocity cone $V^{u_k}(\chi_k)$, resp. $V^{u_k}(P_k)$, in the sense of the above mentioned exponential fall-off properties of $\mathcal{P}_k(R_\tau)$ when the closure of R does not intersect it.

On the other hand, the interaction laws also become analogous, in the $\tau \to \infty$ limit, to those of the classical model, in view of the physical assumption that any energy-momentum transfer which cannot be attributed to (stable) physical particles according to the classical laws gives effects that are damped exponentially with distance, (i.e. with the space-time dilation parameter τ).

Semi-classical arguments[10] then lead to the following first statement of the macrocausality property :

Macrocausality (I)

Let K be any given compact set in $\mathcal{U}$-space of displacements $\mathcal{U} = \{u_k\}$ which are all non causal at $p = \{p_k\}$, (in the sense of section I), whenever each p_k lies in the support of χ_k, resp. are non causal at $P = \{p_k\}$. Then the transition probability W for the scattering process has the following bounds for all positive integers N and all sufficiently small γ $(0 \leq \gamma \leq \gamma_K ,\ \gamma_K > 0)$:

$$\left(\prod_k \|\varphi_{k\tau}\|^2 \right) W(\{\varphi_{k\tau}^{\tau u_k}\}) < \frac{C'_N}{1 + \tau^N} e^{-\alpha'\gamma\tau} , \quad \alpha' > 0 , \tag{12}$$

resp., has the bounds :

$$\left(\prod_k \|\varphi_{k\tau}\|^2 \right) W(\{\varphi_{k\tau}^{\tau u_k}\}) < C' e^{-\alpha'\gamma\tau} , \quad \alpha' > 0 , \tag{12'}$$

where C'_N, C', α', γ_K may depend on K and $\{\chi_k\}$, but are independent of γ and τ.

Remarks :

1) The left-hand side of (12)(12') is $|S_{IJ}(\{\varphi_{k\tau}^{\tau u_k}\ (*)\})|^2$ (see Eq.(2)).

2) In view of the bound (3) ($W \leq 1$), it turns out[2] that the bound (12') can be derived from the bound (12) (for a different set of functions χ_k with smaller supports around P_k).

Conversely, it can also be proved that (12') implies (12)[12].

Therefore either one of the two bounds may be removed from the above statement.

Now, if $\mathcal{U}$ is causal, but is not connected at $p = \{p_k\}$, (resp. at $P = \{P_k\}$), i.e. corresponds to a causal configuration which is composed of several disconnected parts linking the external (initial and final) particles of various subsets $(\bar{I},\bar{J})$ of (I,J), then the physical ideas discussed earlier lead to the further requirement that the remainder $W - \Pi\, W_{\bar{I}\bar{J}}$ should again satisfy bounds of the form (12)(12').

It is proved in the second paper of [1] that a slight generalization of this property (which involves the same physical ideas) yields an analogous property, without phases, for

$$S_{IJ}(\{\varphi_{k\tau}^{\tau u_k(*)}\}) - \prod_{(\bar{I},\bar{J})} S_{\bar{I}\bar{J}}(\{\varphi_{k\tau}^{\tau u_k(*)}\}_{k\in(\bar{I},\bar{J})}) \quad ,$$

and that the latter, yields (if a few exceptional situations are excluded) the following bounds for the connected amplitudes (defined in the Introduction) :

Macrocausality (II)

When there is no connected causal $\mathcal{U}$ in K at $p = \{p_k\}$, with p_k in the support of χ_k, resp. at $P = \{P_k\}$, then the following bounds are satisfied for all sufficiently small γ $(0 \leq \gamma \leq \gamma_K, \gamma_K > 0)$:

$$|S_{IJ}^c(\{\varphi_{k\tau}^{\tau u_k(*)}\})| < \frac{C_N}{1+\tau^N} e^{-\alpha\gamma\tau} \qquad (13)$$

resp. $$|S_{IJ}^c(\{\varphi_{k\tau}^{\tau u_k(*)}\})| < C\, e^{-\alpha\gamma\tau} \quad , \qquad (13')$$

where C_N, C and α are independent of γ (and τ).

The bound (13') can again be derived from (13), and conversely.

Macrocausality-II is slightly stronger than macrocausality-I, and in fact implies it, as easily seen.

Essential support properties

The bounds (12) and (13) directly provide the following essential support properties :

The essential support of $S_{(IJ)}$, resp. T_{IJ}, at a point $P = \{P_k\}$ of $M_{(IJ)}$, is (contained in) the set of causal directions $\tilde{\mathcal{U}}$ at P (in the sense of section I), resp. is (contained in) the set of causal directions $\tilde{\mathcal{U}}$ at P which correspond to connected configurations.

To see this, we note that $S_{IJ}^c(\{\varphi_{k\tau}^{\tau u_k(*)}\})$ (for instance) can be written in the form(*) :

$$\int T_{IJ}(p)\ \hat{\chi}(p)\ e^{-\gamma\tau\Phi(p)}\ e^{i\langle\tau\, \mathcal{U},p\rangle}\ \times$$

$$\times \left[\delta^4 \left(\sum_{i\in I} p_i - \sum_{j\in J} p_j \right) \prod_k \delta(p_k^2 - m_k^2) \right] d^4 p_k \ , \qquad (14)$$

where $p = \{p_k\}$, $\hat{\chi}(p) = \prod_{i\in I} \chi_i^*(p_i) \prod_{j\in J} \chi_j(p_j)$, $\Phi(p) = \sum_k (\vec{p}_k - \vec{\underline{p}}_k)^2$.

Since functions $\hat{\chi}$ with arbitrarily small supports around $\underline{p}$ may be chosen (by using functions χ_k with sufficiently small supports), the above essential property for T_{IJ} directly follows from the definitions of Appendix I-B (see also lemma 4 at the end of section E).

III - PHYSICAL-REGION ANALYTICITY PROPERTIES

In view of Theorems 1 and 2 of Appendix I-C, the essential support property of T_{IJ} and the results on causal displacements given in paragraphs a)b)c)d) at the end of part I, directly provide the following analyticity properties :

a) If $p = \{p_k\}$ is not a $+\alpha$ - Landau point, then T_{IJ} is analytic at p .

b) In the neighborhood of a $+\alpha$ - Landau point p which lies on one surface $L_0^+(G)$, T_{IJ} is the boundary value of an analytic function from well defined "plus $i\varepsilon$" directions.

More precisely, let $q = \{q_\rho\}$, $\rho = 1,\dots 3(|I|+|J|-4)$, be any system of real analytic local coordinates of $M_{(IJ)}$ at p , and let ℓ be a real analytic function of q such that i) $L_0^+(G)$ is locally represented by the equation $\ell(q)=0$, and ii) $\ell(q) > 0$ on the physical side of $L_0^+(G)$. Then, in a neighborhood N of p

$$T_{IJ} = \lim_{\substack{|\eta|\to 0 \\ \eta\,\in\, C}} F(q + i\eta) \qquad (15)$$

in the sense of distributions, where the directions η are those of a cone C

(*) The θ-functions of the energies can be removed from the integrand of (14) in the neighborhood of a given $\underline{p}$ in $M_{(IJ)}$.

strictly contained in the half-space $\nabla \ell_{|p} \cdot \eta > 0$. This cone can be chosen arbitrarily close to this half-space if $\mathcal{N}$ is chosen sufficiently small.

c) In the situation of paragraph c) at the end of section I, the same result holds with the half-space $\nabla \ell . \eta > 0$ being now replaced by the (non empty) intersection of the half-spaces associated with the various contractions G' of G.

Since the remaining points belong to lower-dimensional submanifolds of $M_{(IJ)}$, the above results immediately provide (theorem 2 of Appendix I) :

Property 1

For each given physical process $I \to J$, there exists a unique analytic function F_{IJ} (*) to which T_{IJ} is equal at all non $+\alpha$ -Landau points, and from which it is a "plus iε" boundary value at all the above mentioned $+\alpha$ - Landau points.

d) Finally, in the neighborhood of a point p which lies at the intersection of several $+\alpha$ -Landau surfaces with no common parent, T_{IJ} can be decomposed as a sum of boundary values f_β (in the sense of distributions) of analytic F_β , each of which from the directions of a cone C_β associated with one of the graphs, or parent graphs, involved at p .

Independence property

When p is a "type I" point (i.e. when the various cones $\tilde{C}_\beta$ of causal directions at p are disjoint), it was moreover (partly) proved in [1] that each f_β "has the same analytic properties, in the neighborhood of p, as if no other parent surface was involved", namely that i) f_β is analytic at all real points which do not belong to the corresponding parent surface $L^+(G_\beta)$ or to the surfaces $L_o^+(G'_\beta)$ of the "contractions" G'_β of G_β involved at p, and ii) that it obeys the "plus iε" rule of paragraphs b) and c) at these points.

When the cones $\tilde{C}_\beta$ are disjoint, this is in fact a direct consequence of properties a) to d) above (see Appendix I-D).

Consider for instance a point p' (of the neighborhood of p) which lies on no $+\alpha$ -Landau surface, and at which the essential support of T_{IJ} is therefore empty (T_{IJ} is analytic). From property d), the essential support of f_β

(*) Defined in a domain of the complexified manifold $\overline{M}_{(IJ)}$ of $M_{(IJ)}$.

at p' is contained in the dual cone $\tilde{C}_\beta$ of C_β .On the other hand, since :

$$f_\beta = T_{IJ} - \sum_{\beta'\neq\beta}' f_{\beta'}$$

it is also contained in the union, for $\beta' \neq \beta$, of the dual cones $\tilde{C}_{\beta'}$ of $C_{\beta'}$. Since the sets $\tilde{C}_\beta$ and $\bigcup_{\beta'\neq\beta} \tilde{C}_{\beta'}$ are disjoint (for N sufficiently small), the essential support at p' is therefore empty, and f_β is analytic there.

In view of theorem 3 of Appendix I , this "independence property" also follows at type II points. Moreover, as a result of this theorem, global decomposition properties of T_{IJ} into sums of boundary values (in the sense of distributions) of analytic functions can be derived in $M_{(IJ)}$. The analytic functions involved are independent of p in $M_{(IJ)}$, although the directions η from which their boundary value is obtained may depend on it. They are moreover analytic at all real points outside given $+\alpha$ - Landau surfaces.

A detailed analysis of these decomposition properties, which are the generalization of the above property 1), when the points p of paragraph d) are included, will be given elsewhere.

REFERENCES

[1] - D. IAGOLNITZER and H.P. STAPP ; Comm. Math. Phys. **14**, 15 (1969)
Further related results, also obtained in collaboration with Dr. H.P.STAPP, have been given in

D. IAGOLNITZER ; in Lectures in Theoretical Physics, ed. by K. Mahanthappa and W. Brittin, Gordon and Breach, New-York (1969), p.221.

[2] - D. IAGOLNITZER ; Introduction to S-matrix theory, A.D.T. 21 rue Olivier-Noyer, Paris 75014, France (1973).

[3] - J. BROS and D. IAGOLNITZER ; Local Analytic Structure of Distributions ; I - Generalized Fourier transformation and essential supports (In this volume).

The results of this work, which are also contained in a large extent in Ch.II.C of reference 2, follow from an elaboration of the previous results of reference 1 and of :

J. BROS and D. IAGOLNITZER ; in Proceedings of the 1971 Marseille meeting on Renormalization Theory, and Ann. Inst. Poincaré, Vol. 18, n°2 (1973),p147.

[4] - J. BROS and D. IAGOLNITZER ; Local Analytic Structure of distributions; II - General decomposition theorems.

The results of this work which are extensions of those of references 1 and 3 are also very closely linked to, and have been inspired to a large extent by, the results of reference 5 below :

[5] - M. SATO, T. KAWAÏ, M. KASHIWARA ; in "Hyperfunctions and Pseudo-Differential equations" - Lecture Notes in Mathematics,Springer-Verlag, Heidelberg (1973).

The basic results of Professor M. Sato and his coworkers have been presented in the course by :

J. CHAZARAIN, A. CEREZO, A. PIRIOU (In this volume)

[6] - F. PHAM ; Micro-analyticité de la matrice S (in this volume).

[7] - The statement of the macrocausality property given here is essentially that of reference 2, which is a slight improvement of that given in reference 1.

The latter was itself inspired from previous macrocausality properties presented in various situations in :

C. CHANDLER and H.P. STAPP ; J. Math. Phys. 90, 826 (1969)
R. OMNES ; Phys. Rev. 146, 1123 (1966)
F. PHAM ; Ann. Inst. Henri Poincaré, Vol.6, n°2, 89 (1967)

Earlier pioneer works on macrocausality are due to :

G. WANDERS ; Nuov. Cim. 14, 168 (1959) and Helv. Phys. Acta 38, 142 (1965).

On the present physical status of macrocausality, the interested reader may also be referred to :

H.P. STAPP ; S-matrix theory, in preparation and to Lawrence Berkeley Laboratory reports by the same author, in particular. "Macrocausality and its role in Physical Theories", and "Foundations of S-matrix theory".

[8] - The interest of this classical model and the possibility of its relevance to quantum relativistic physics have been for instance emphasized by

C. COLEMAN and R. NORTON ; Nuov. Cim. 38, 438 (1965)

who have noticed that the $+\alpha$ - Landau equations are equations of classical kinematics.

The basic results on the $+\alpha$ - Landau surfaces and the causal displacements at $+\alpha$ - Landau points are due to

C. CHANDLER and H.P. STAPP, op.cit.
F. PHAM, op. cit.

A detailed presentation of the model, with simple examples, will also be found in Ch.II-B of reference 2

[9] - C. CHANDLER ; Helv. Phys. Acta 42, 759 (1969)

[10] - A more detailed physical discussion is given in references 1 and 2 . See also further aspects of this discussion in the works by Dr. H.P. STAPP mentioned at the end of reference 7.

[11] - Microlocal essential support of a distribution and decomposition theorems, an Introduction, Saclay preprint, and in this volume.
The variables p and τu correspond to the variables x and ξ respectively of this mathematical Appendix.

[12] - This follows from results proved in J. BROS and D. IAGOLNITZER (in preparation).

APPENDIX

MICROLOCAL ESSENTIAL SUPPORT OF A DISTRIBUTION AND DECOMPOSITION THEOREMS - AN INTRODUCTION

D. IAGOLNITZER
DPh-T CEN Saclay BP n°2 91190 Gif-sur-Yvette
FRANCE

This text is a short mathematical introduction to the notion of essential support of a distribution in the "microlocal" sense and to the corresponding decomposition theorems of distributions into sums of boundary values of analytic functions. Results concerning the multiplication of distributions and their restrictions to submanifolds are also mentioned at the end. The details are given in references 1, 2 (and in the references quoted therein), in which a more general "non microlocal" notion of essential support and corresponding results are also given.

It can be checked (see section C) that the "microlocal" essential support coincides with the "singular support" or "singular spectrum" introduced independently by very different methods (and for general hyperfunctions) by Professors SATO, KAWAI, KASHIWARA [3][4](*).

A - GENERALIZED FOURIER TRANSFORMATION

Let f be a tempered distribution defined in the n-dimensional real vector space $R^n_{(x)}$ of a variable $x = (x_1, \dots x_n)$. For our purposes below, we assume for simplicity that f has a compact support.

The generalized Fourier transform $W_o(f)$ at a point x_o (of $R^n_{(x)}$) is defined in the real space $R^{n+1}_{(\xi,\xi_o)}$ of the variables $\xi = \xi_1, \dots \xi_n$ and of a supplementary variable ξ_o by the formula :

$$\{W_o(f)\}(\xi,\xi_o) = \int f(x)\, e^{-i\xi . x - \xi_o \Phi(x-x_o)} dx \quad , \qquad (1)$$

(*) It may also coincide with the "analytic wave front set" introduced in [5], although the corresponding analyticity properties have not so far been exhibited in general, to our knowledge, in this latter case.

where $\Phi(x) = x^2 = \sum_{i=1}^{n} x_i^2$.

The function Φ may also be chosen to be an other function with analogous local properties in the neighborhood of $x = 0$: Φ is analytic, $\Phi(z^*) = (\Phi(z))^*$ where $z = x + iy = (z_1, \ldots z_n)$ and $*$ is complex conjugation, $\Phi(0) = 0$ and Φ has a critical point $(\nabla\Phi(0) = 0)$ "with positive signature" at $x = 0$. For simplicity, the reader may keep in mind the function $\Phi = x^2$ and the dependence of W_0 on Φ will be left implicit.

We note that $\{W_0(f)\}(\xi, 0)$ is the usual Fourier transform of f . The first following lemma holds :

Lemma 1 : Let g be a C^∞ function with compact support around x_0 . Then g is locally analytic at x_0 if and only if its generalized Fourier transform $W_0(g)$ at x_0 satisfies the following bounds with uniform constants $\alpha > 0$, $\gamma_0 > 0$, and $C_N < \infty$:

$$|W_0(g)\,(\xi, \gamma|\xi|)| < \frac{C_N}{1+|\xi|^N}\, e^{-\alpha\gamma|\xi|} , \qquad (2)$$

for all positive γ less than γ_0 , and all positive integers N .

The bounds (2) at $\gamma = 0$ express the C^∞ character of g . It is well known that the local analyticity at x_0 cannot be expressed in terms of an exponential decrease of the usual Fourier transform of g . Lemma 1 tells us that it can be characterized in terms of the above exponential decrease of the generalized Fourier transform.

Proof : a) The proof that local analyticity implies (2) is obtained by distorting the part $0 \leq \Phi\,(x-x_0) \leq \alpha$ of the integration domain in (1) (with α sufficiently small) in the analyticity domain of $g^{(*)}$.

For $N > 0$, appropriate integrations by parts are also used.

b) For the converse proof, it is useful to consider, together with $W_0(g)$, a set of n functions $W_k(g)(k = 1, \ldots n)$ of ξ, ξ_0 , and z , defined by

---

(*) Use a distorted contour lying of the surface

$$y + b\,(\mathrm{Re}\ \Phi(x - x_0 + iy) - \alpha) = 0$$

with b appropriately chosen (for each direction of ξ) and $|b|$ sufficiently small.

$$\{W_k(g)\}(\xi,\xi_o,z) = i\int g(x')\, e^{-i\xi.x' - \xi_o\Phi(x'-x_o)}\, \rho_k(x'-x_o,z-x_o)\, dx' \quad (3)$$

where $\rho_k(z,z') = z_k + z'_k$ if $\Phi = z^2$; more generally ρ_k is an analogous (locally analytic) function such that

$$\Phi(z) - \Phi(z') = \sum_{k=1}^{n} \rho_k(z,z')\,(z_k - z'_k)\,. \quad (4)$$

In view of this property, one checks that the following differential form of degree n in (ξ,ξ_o) - space :

$$W_x(g) = e^{i\xi.x + \xi_o\Phi(x-x_o)} \sum_{k=0}^{n} (-1)^k\, W_k(g)(\xi,\xi_o,x)\, d\xi_o\wedge\ldots d\hat{\xi}_k\wedge\ldots d\xi_n \quad (5)$$

where the notation $d\hat{\xi}_k$ means that this factor is omitted, is <u>closed</u> ($dW_x = 0$) for each value of x . On the other hand :

$$\int_{\xi_o=0} W_x(g) \left(= \int W_o(g)(\xi,0)\, e^{i\xi.x}\, dx\right) = (2\pi)^{n/2}\, g(x). \quad (6)$$

We show below that all functions $W_k(g)$ satisfy bounds of the form (2) if $W_o(g)$ does, with constants C_N, α, γ_o which can be chosen independent of $k = 0,\ldots n$ and of z in any bounded complex neighborhood N of $x = 0$(*).

In view of these bounds and of Stokes theorem, the integration surface $\xi_o = 0$ in (6) can be distorted on the surface $\xi_o = \gamma_o|\xi|$ when $\Phi(x-x_o) < \alpha$, and the local analyticity of g at x_o is in turn derived from the exponential bounds of the functions W_k on this surface.

Finally, the above result on $W_k(g)$ is proved by considering it as the Fourier transform (see (3)) of $g\, e^{-\xi_o(1-\eta)\Phi} \times \rho_k\, \varphi\, e^{-\xi_o\eta\Phi}$ where $0 < \eta < 1$ and φ is a C^∞ function with compact support, which is equal to one in the support of g . Correspondingly :

$$\{W_k(g)\}(\xi,\xi_o,z) = \int d\xi'\; W_o(g)(\xi-\xi',(1-\eta)\,\xi_o) \times W_o(\rho_k\varphi)\,(\xi',\eta\,\xi_o) \quad . \quad (7)$$

(*) Some restrictions on the width of N may have to be added for a general function Φ (with no inconvenience for the purposes of this text).

The announced result is obtained from the assumed bounds on $W_o(g)$ and from the fact that $W_o(\rho_k\varphi)$ satisfies analogous bounds (which are shown to be independent of z in N) in view of the local analyticity of $\rho_k\varphi$ at x_o with respect to x' (see part a) of the proof). To see this, divide for instance the integration domain in (7) into two parts : $|\xi'| < \varepsilon\ |\xi|, |\xi'| > \varepsilon\ |\xi|$.

B - ESSENTIAL SUPPORT OF A DISTRIBUTION (possibly defined on a manifold).

Being given a distribution f defined on $\mathbb{R}^n_{(x)}$, and C^∞ functions χ with compact support around x_o , locally analytic and different from zero at x_o , we denote by $\Sigma_\chi(f)$ the set of directions $\hat{\xi}$ in $R^n_{(\xi)}$ along which $W_o(\chi f)$ does not satisfy bounds of the type (2) : namely, a direction $\hat{\xi}_o$ is in the complement of $\Sigma_\chi(f)$ if there exists a neighborhood U of $\hat{\xi}_o$, constants $\alpha > 0$, $\gamma_o > 0$, and $C_N < \infty$ such that the bounds (2) be satisfied by $|W_o(\chi f)|$ for all points $\xi = \lambda\hat{\xi}$, $\hat{\xi} \in U$, $\lambda \geq 0$.

The following lemma holds :

Lemma 2 : $\Sigma_{\chi_2}(f) \subset \Sigma_{\chi_1}(f)$ whenever χ_2 has its support in the region where $\chi_1 \neq 0$ around x_o .

(This is proved by writing $\chi_2 f = \chi_1 f \times \chi_2\chi_1^{-1}$ and using a convolution argument analogous to that used at the end of section A).

Definition : The essential support $\Sigma_{x_o}(f)$ of f at x_o , is the limit of $\Sigma_\chi(f)$ when the width $|\text{supp}\ \chi|$ of the support of χ around x_o tends to zero.

In view of lemma 2, this definition is clearly independent of the sequence of functions χ considered. Theorem 1 of section C also shows that it is independent of the function Φ considered , since the analyticity properties involved there do not depend on Φ .

If f is defined on a manifold M rather than on $\mathbb{R}^n$, $\Sigma_{x_o}(f)$ is defined as above for any system of local coordinates at x_o . Theorem 1 of section C again ensures that it is a well defined subset of the cotangent vector space $T^*_{x_o}M$ at x_o to M , independent of the choice of local coordinates.

The essential support (in the "microlocal" sense) $\Sigma(f)$ of f is the closed subset of the (sphere) cotangent bundle T^*M defined as

$$\bigcup_{x \in M} x \times \Sigma_x(f) \ . \tag{8}$$

<u>Remark</u> : $\Sigma_{x_o}(f)$ has been defined above as a subset of directions $\hat{\xi}$. We also identify it (sometimes) with the corresponding cone in $R^n_{(\xi)}$ with apex at the origin, from which the origin is removed.

C - A LOCAL DECOMPOSITION THEOREM

We first fix some notations : $\tilde{C}_\beta$ will denote an open convex salient cone in ξ-space with apex at the origin , and C_β will denote its open dual cone in y-space. Then the following theorem, which is an extension of lemma 1 holds :

<u>Theorem 1</u> : There is equivalence between the two following properties

i) $\Sigma_{x_o}(f)$ is contained in the union of a finite family of cones $\tilde{C}_\beta$.

ii) There exists a neighborhood $\mathcal{U}$ of x_o, and boundary values f_β in $\mathcal{U}$ (in the sense of distributions) of analytic functions F_β, from the respective directions y of the cones $\bar{C}_\beta^{(*)}$, such that :

$$f = \sum_\beta f_\beta \quad \text{in } \mathcal{U} .$$

<u>Proof</u> : a) The proof that ii) implies i) is obtained by introducing the function :

$$h(x) = e^{\frac{1}{\Phi(x-x_o)-\alpha}} \qquad \text{when } 0 \leq \Phi(x-x_o) \leq \alpha$$

$$= 0 \qquad \text{outside this region} \tag{9}$$

with α such that the region $0 \leq \Phi(x-x_o) < \alpha$ lies in $\mathcal{U}$. The fact that $\Sigma_h(f_\beta)$ is contained in $\tilde{C}_\beta$, is obtained as in lemma 1 by distorting this region in the analyticity domain of F_β (and of h)$^{(**)}$.

b) For the converse proof, we introduce a C^∞ function χ, locally analytic and different from zero at x_o, with support such that $W_o(\chi f)$ satisfies the bounds (2) outside $\bigcup_\beta \tilde{C}_\beta$ with uniform constants C_N, α, γ_o. As in lemma 1,

---

(*) $\bar{C}_\beta$ is the closure of C_β ; i.e. the directions y are those of open cones "slightly larger" than C_β.

(**) h is analytic in a complex (open) neighborhood of $0 \leq \Phi(x-x_o) < \alpha$, and one checks that hF, as well as its derivatives, is bounded on the distorted contours, in view of the properties of h.

it can be shown that all functions $W_k(\chi f)$ also satisfy analogous uniform bounds outside $\bigcup_\beta \tilde{C}_\beta$. We now consider, as in Eq.(6), the equality :

$$\chi f(x) = \int_{\xi_o = 0} W_x(\chi f) \quad . \tag{10}$$

The contribution to this integral of the complement of $\bigcup_\beta \tilde{C}_\beta$ in the hyperplane $\xi_o = 0$ is again treated by distorting this domain in the region $\xi_o > 0$. It provides a C^∞ function which is a sum of boundary values (in the sense of functions) of analytic functions $F_{1\beta}$ from the directions y of the cones C_β, when x is in the neighborhood of x_o(*). The contribution of $\bigcup_\beta \hat{C}_\beta$ in the hyperplane $\xi_o = 0$ is treated by the usual Laplace transform theorem and easily provides a sum of boundary values (in the sense of distributions) of analytic functions $F_{2\beta}$, again from the directions y of the cones C_β. The result is therefore proved.(**)

<u>Remarks</u> : 1)[1] Being given a function Φ, a given point x_o, a given $\alpha > 0$ and the corresponding set Ω_α $(0 \leq \Phi(x-x_o) < \alpha)$, a notion of Ω_α-essential support in (ξ,ξ_o)-space, which generalizes the set $\Sigma_{x_o}(f)$ in ξ-space, can be introduced by methods which are similar to those used above. Theorem 1 then appears as a limit case of a more general decomposition theorem, which provides corresponding decompositions of f <u>in Ω_α</u> into sums of boundary values of analytic functions whose analyticity domains can be specified. The cones C_β in y-space from which the boundary values are obtained are here independent of x in Ω_α.

A corollary of this result is the usual generalized edge-of-the wedge theorem, over any bounded set Ω (with smooth boundary), with possibly precise specifications of the analyticity domains involved (see ref.1).

2) According to a definition given in [4], f is said to be "micro-analytic" at a point $(x_o,\hat{\xi}_o)$, or $(x_o,\hat{\xi}_o)$ is said to be in the complement of the "singular spectrum" of f, if f can be locally decomposed, in the neighborhood of x_o, as a sum of boundary values $\hat{f}_\beta$ of analytic functions $\hat{F}_\beta$ from directions y which all lie in the region $\hat{\xi}_o . y < 0$.

(*) In contrast to lemma 1, this contribution is not analytic, because of the parts of the surface joining the boundary of $\bigcup_\beta \tilde{C}_\beta$ at $\xi_o = 0$, to the surface $\xi_o = \gamma|\xi|$.

(**) To obtain cones "slightly larger" than the cones C_β, replace the cones $\check{C}_\beta$ by "slightly smaller" cones whose union still contains the closed set $\Sigma_{x_o}(f)$.

The part ii) → i) of theorem 1 ensures, as easily checked, that $\hat{\xi}_o$ is then also in the complement of the "essential support" $\Sigma_{x_o}(f)$ of f at x_o.

Theorem 1 then provides the following lemma :

Lemma 3 : Being given any compact set K of directions $\hat{\xi}$ such that f is micro-analytic at $(x_o,\hat{\xi})$ for all $\hat{\xi}$ in K (in the sense of the above definition), f can be decomposed, in the neighborhood of x_o, as a sum of boundary values f_β of analytic functions F_β, which are independent of $\hat{\xi}$ in K(*).

D - GENERAL (MICROLOCAL) DECOMPOSITION THEOREM

Being given a distribution f with essential support $\Sigma(f)$, theorem 1 provides decompositions of f into sums of boundary values of analytic functions at each point x_o, but gives so far no information on the links between decompositions obtained at different points. Theorems 2 and 3 below provide this information in various situations.

Theorem 2 below is a direct corollary of theorem 1 (in view of arguments of analytic continuation).

Theorem 2 : The following properties are equivalent :

i) the fiber $\Sigma_x(f)$ of $\Sigma(f)$ at all points x of an open set Ω is contained in a closed convex salient cone $\tilde{C}_x$ (which may be depend on x and may be empty at some points).

ii) f is the boundary value, in Ω, of a unique analytic function F from directions y which are those of the dual cone C_x of $\tilde{C}_x$, at each point x (F is analytic at x, if $\tilde{C}_x$ is empty).

We next state :

Theorem 3 : Let $\Sigma(f)$ be contained in the union of a finite family of closed subsets Σ_β of $T^*\Omega$, whose fibers $(\Sigma_\beta)_x$ are closed convex salient cones for all values of β and x (in Ω).

(*) For any $\hat{\xi}_o$ in K, each boundary value f_β can still be obtained from directions y such that $\hat{\xi}_o \cdot y < 0$.

Then f is equal, in Ω, to a sum of distributions f_β whose essential support $\Sigma(f_\beta)$ is contained in Σ_β, for each value of β. According to theorem 2, each f_β is the boundary value of a corresponding analytic function F_β.

We give an idea of the proof below in the case when the fibers $(\Sigma_\beta)_x$ are <u>disjoint</u> for every point x in Ω $((\Sigma_\beta)_x \cap (\Sigma_{\beta'})_x = \emptyset \ \forall\ \beta,\beta',\beta \neq \beta'$, $\forall x \in \Omega)$.

Consider a given point x_o in Ω and a family of disjoint cones $\tilde{C}_\beta$ containing the respective sets $(\Sigma_\beta)_{x_o}$. Theorem 1 ensures the existence of a neighborhood $\mathcal{U}$ of x_o where

$$f = \sum_\beta f_\beta \quad ,$$

with $\Sigma_x(f_\beta) \subset \tilde{C}_\beta$, $\forall\beta$, $\forall x \in \mathcal{U}$.

We now show that $\Sigma_x(f_\beta) \subset (\Sigma_\beta)_x$: being given any family of (disjoint) $\tilde{C}'_\beta$ containing the sets $(\Sigma_\beta)_x$, theorem 1 again ensures the existence of a neighborhood $U \subset \mathcal{U}$ of x such that :

$$f = \sum_\beta g_\beta \qquad \text{in } U \quad ,$$

with $\Sigma_x(g_\beta) \subset \tilde{C}'_\beta$.

Therefore, one has :

$$\sum_\beta (f_\beta - g_\beta) = 0 \qquad \text{in } U \quad ,$$

and since the cones $\tilde{C}_\beta$ are disjoint, it clearly follows that $\Sigma_x(f_\beta - g_\beta)$ is empty, for every β. Hence :

$$\Sigma_x(f_\beta) \subset \quad \Sigma_x(g_\beta) \subset \tilde{C}'_\beta \quad , \ \forall\beta \quad .$$

Since the cones $\tilde{C}_\beta$ may be chosen arbitrarily close to the sets $(\Sigma_\beta)_x$, theorem 3 is proved in $\mathcal{U}$.

The proof of theorem 3 in Ω then results from an application of Cousin theorem (see part V reference 1).

When the cones $(\Sigma_\beta)_x$ are no longer disjoint, the proof is somewhat more complicated and we refer the interested reader to [2].

Remark : Theorem 3 is very closely linked with basic results of reference 3 on the "sheaf of microfunctions". It seems possible[6] to derive somewhat more refined results of the latter type, which, in particular, also hold in the framework of distributions, by making use of the generalized Fourier transformation, together with an application of Cousin theorem which is adapted from [3].

E - MULTIPLICATION OF DISTRIBUTIONS AND RESTRICTIONS TO SUBMANIFOLDS[7]

a) Product of two distributions

Theorem 4 : A sufficient condition for the product $f_1 f_2$ of two distributions to be well defined is

$$\Sigma(f_1) \cap \left(- \Sigma(f_2)\right) = \emptyset \quad . \tag{12}$$

The fiber $\Sigma_x(f_1 f_2)$ of $\Sigma(f_1 f_2)$ at x is contained in the set $\Sigma_x(f_1) + \Sigma_x(f_2)$ of vectors ξ of the form $\xi_1 + \xi_2$, $\xi_1 \in \Sigma_x(f_1)$, $\xi_2 \in \Sigma_x(f_2)$.

Proof : This problem is clearly a local problem in the neighborhood of each point x. It is easily solved by using theorem 1 and the natural definition of the product of two distributions which are boundary values of analytic functions from common directions (see [1]).

Alternatively, the product $\chi_1 f_1 \times \chi_2 f_2$, where χ_1, χ_2 are C^∞ functions, locally analytic and different from zero at x, and with sufficiently small support around x is defined as the inverse Fourier formula of :

$$\int (\widetilde{\chi_1 f_1})(\xi') \, (\widetilde{\chi_2 f_2})\,(\xi-\xi')\, d\xi' \tag{13}$$

where $\widetilde{\chi_i f_i}$ is the Fourier transform of $\chi_i f_i$.

In view of the slow increase of $(\widetilde{\chi_1 f_1})$ and $(\widetilde{\chi_2 f_2})$ and of their rapid decrease outside $\Sigma_{\chi_1}(f_1)$ and $\Sigma_{\chi_2}(f_2)$ respectively (bounds (2) at $\gamma = 0$), this integral is in fact convergent for all ξ, and defines a slowly increasing function of ξ, whose Fourier transform is therefore a well defined distribution. To check that $W_o(\chi_1 f_1 \times \chi_2 f_2)$ does satisfy bounds of the type (2) in all directions $\hat{\xi}$ which do not belong to $\Sigma_{\chi_1}(f_1) + \Sigma_{\chi_2}(f_2)$, write (in the same may as in (6)) :

$$\{W_o(\chi_1 f_1 \times \chi_2 f_2)\}(\xi,\xi_o) = \int d\xi' \, \{W_o(\chi_1 f_1)\}(\xi',(1-\eta)\xi_o) \times \{W_o(\chi_2 f_2)\}(\xi-\xi',\eta\,\xi_o) \quad (14)$$

where $0 < \eta < 1$, and use the bounds (2) for $W_o(\chi_1 f_1)$ and $W_o(\chi_2 f_2)$ outside $\Sigma_{\chi_1}(f_1)$ and $\Sigma_{\chi_2}(f_2)$: the result is ensured by the fact that the intersection of neighborhoods (in ξ'-space) containing the cones $\Sigma_{\chi_1}(f_1)$ (with apex at the origin) and $\xi-\Sigma_{\chi_2}(f_2)$ (with apex at ξ) is empty.

The announced property of $\Sigma_x(f_1 f_2)$ follows from the fact that χ_1, χ_2 can be chosen with arbitrarily small supports(*).

b) <u>Restriction of a distribution to a submanifold and related results</u>

<u>Theorem 5</u> : Let M be a submanifold of on open set Ω of R^n and let N be the conormal bundle to M. A sufficient condition for a distribution f on Ω to have a well defined restriction $f_{|M}$ to M (as a distribution) is :

$$\Sigma(f) \cap N = \emptyset \; . \quad (15)$$

The fiber $\Sigma_x(f_{|M})$ of this restriction at a point x of M is contained in the quotient $\Sigma_x(f)/N_x$ of $\Sigma_x(f)$ by the vector space N_x conormal at x to M.

<u>Proof</u> : It is also sufficient here to solve the problem locally, in the neighborhood of each point x_o . As in subsection a) the proof may be derived from theorem 1 (see [1]). Alternatively, let M be defined in the neighborhood of x_o by a set of ℓ équations $L_i(x) = 0$ $(i = 1,\ldots,\ell)$ where each L_i is a real analytic function of x , and let $\delta(M) = \prod_{i=1}^{\ell} \delta(L_i(x))$. It can be checked that this is (locally) a well defined distribution whose essential support at x_o is N_{x_o} . According to theorem 4 and to (15), $\delta(M)\times f$ is therefore (locally) a well defined distribution, and $\Sigma_{x_o}(\delta(M)\times f) \subset \Sigma_{x_o}(f) + N_{x_o}$.

The definition of $f_{|M}$ follows. To see that $\Sigma_{x_o}(f_{|M}) \subset \Sigma_{x_o}(f)/N_{x_o}$, one may for instance choose a set $\{q\}$ of $n-\ell$ local coordinate of M at x_o among the set of n variables $x_1 \ldots x_n$, and consider the subspace Γ of $\mathbb{R}^n_{(\xi)}$ of

---

(*) It is easy to check that the definition of $f_1 f_2$ "at x" does not depend on the choice of χ_1 and χ_2 .

vectors whose ℓ components, corresponding to those excluded above, are fixed at zero. It is easily seen that $\Gamma \cap N_{x_o} = 0$ and that there is a well defined 1-1 correspondence between Γ and $\mathbb{R}^n_{(\xi)}/N_{x_o}$. The announced result then easily follows from the definition of the essential support at x_o of distributions defined in $\mathbb{R}^{n-\ell}_{(q)}$, the needed bounds (2) being ensured by the corresponding bounds on $W_o(f\delta(M))$(*).

The last part of the proof also provides the following lemma :

<u>Lemma 4</u> : Let f be a distribution <u>defined on M</u>, $f\delta(M)$ be its product with $\delta(M)$ (defined in $\mathbb{R}^n_{(x)}$) and $\Sigma_{x_o}(f\delta(M))$ its essential support at x_o ;Then $\Sigma_{x_o}(f\delta(M))$ be invariant by addition of vectors in N_{x_o}, and $\Sigma_{x_o}(f)$ is the quotient $\Sigma_{x_o}(f\delta(M))/N_{x_o}$.

(*) Note that the restriction of a function Φ defined in $\mathbb{R}^n_{(x)}$ to M provides a function Φ in $\mathbb{R}^{n-\ell}_{(q)}$ which belongs there to the class described below (Eq.1).

REFERENCES

[1] - J. BROS and D. IAGOLNITZER ; Local Analytic Structure of Distributions I - Generalized Fourier transformation and essential supports (in this volume). The method of the generalized Fourier transformation was first introduced in :

D. IAGOLNITZER and H.P. STAPP ; Comm. Math. Phys. 14, 15 (1969)

where lemmas 1 and 4, and theorems 1 and 2 of the present text were essentially proved, in a less elaborate form(and in the framework of S-matrix theory).

Further developments have been given in :

J. BROS and D. IAGOLNITZER ; in Proceedings of the 1971 Marseille meeting on Renormalization Theory, and Ann. Inst. Poincaré, Vol. 18, n°2 (1973)p.147

J. BROS ; in Comptes-Rendus de la R.C.P. 25, Nov.1971, Strasbourg, CNRS
D. IAGOLNITZER, in Ch.II-C of Introduction to S-matrix theory, A.D.T., 21 rue Olivier-Noyer, Paris 75014, France (1973).

[2] - J. BROS and D. IAGOLNITZER ; Local Analytic Structure of distributions ; II - General decomposition theorems.

The results of this work which are extensions of those of references 1 and 3 are also very closely linked to, and have been inspired to a large extent by, the results of reference 3 below :

[3] - M. SATO, T. KAWAÏ, M. KASHIWARA ; in "Hyperfunctions and Pseudo-Differential equations" - Lecture Notes in Mathematics, Springer-Verlag, Heidelberg (1973).

The basic results of Professor M. Sato and his coworkers have been presented in :

[4] - J. CHAZARAIN, A. CEREZO, A. PIRIOU (In this volume)

[5] - L. HÖRMANDER ; Comm. Pure Appl. Math. 24, 671 (1971)

[6] - J. BROS and D. IAGOLNITZER ; in preparation

[7] - Analogous results have been proved in references 3 and 4.
These results can also be extended to the case of "non microlocal" essential supports by using somewhat more refined arguments, J. BROS and D. IAGOLNITZER in preparation.

UNITARITY AND DISCONTINUITY FORMULAE

David OLIVE
CERN, Geneva, Switzerland

1. Introduction

I would like to thank the organizers for enabling me to be here this week in Nice to hear about the new developments concerning hyperfunction theory which seem so well suited to rendering more precise and rigorous certain ideas and results in S-matrix theory which I shall describe. However, first I must apologize for the fact that I have not myself worked on these aspects of S-matrix theory since six years ago, and secondly for the fact that I am not a mathematician and consequently have not fully understood the new mathematical language that I have heard this week.

I want first to make a historical digression to explain my own point of view, which is somewhat more old fashioned than that of Pham [12] and Iagolnitzer [9] in the preceding talks.

Study of the S-matrix *per se* began its activity in the early 1960's [3,17,14, 8,15]. The S-matrix was expected to satisfy many interesting properties, unitarity and analyticity among them, but its detailed structure was not well understood. It was believed by some physicists that the general requirements were so restrictive that there might be a unique solution -- that which describes and contains all useful information about the physical world. Study of these requirements might then lead to this solution. I do not think I believed this then, and probably very few do today, yet in a sense it has become much more plausible with the advent of the dual theory of the S-matrix -- but that is another story, on which I currently work, with its own intriguing brand of mathematics [7,18]. What I did believe then was that the study of these general properties could lead to a more detailed understanding of the structure of many-particle S-matrix elements (which are nowadays observed experimentally) and, in particular, of the singularity structure (in the sense of complex variable theory). You have already heard [12,9] a lot about how this has become true.

The first obvious question was whether the known properties of the S-matrix (known in the sense of having been abstracted from quantum field theory) were independent or interdependent; were there a few basic properties which could be justified by direct physical argument and used to derive the others? It soon became clear [8,15] that the two crucial properties were unitarity and the analytic behaviour of S-matrix elements in terms of the particle momenta for values of these momenta close to the physically permitted ones, i.e. "in the neighbourhood of the physical region". This latter is what we shall now discuss.

2. Postulates of analyticity in the neighbourhood of the physical region

Pham has explained [12] one such postulate which is certainly the most recent and seems to be the most precise, refined, and subtle one, and which in his notation reads

I Microanalyticity .

The scattering amplitude S_{IJ} is microanalytic in the direction of each covector $u \in T^*M_{IJ}$, except for covectors which are causal.

Iagolnitzer has explained another postulate [9], the macrocausality postulate, which is more physical in content and can apparently be shown to be equivalent to the microanalyticity postulate [13,10], i.e. microanalyticity $\Leftrightarrow$ macrocausality.

Ten years ago ignorance, both physical and mathematical, prevented one from making such detailed postulates -- one just did not know or understand the analytic structure of the physical region well enough. Because of this I was forced in my work [15] to make a more vague and less precise postulate:

II $i\varepsilon$ postulate .

The scattering amplitude is analytic in the neighbourhood of all physical points except for those lying on certain curves: the amplitudes defined on either side of these exceptional points are analytically related by paths of continuation which become infinitesimally imaginary. Any closed path of continuation can be contracted to zero since the physical scattering amplitude is single valued.

Actually it is now known that this is not quite correct; one must allow the possibility of a linear decomposition into parts satisfying the above. This illustrates the fact that this $i\varepsilon$ postulate can probably be formulated much more cleanly in terms of hyperfunction language.

Later I shall explain more about another fundamental postulate, one which is universally accepted,

III the unitarity of the S-matrix .

A fourth statement is the spectral assumption of Pham [12] which, he argued, is equivalent to what physicists would call the Cutkosky discontinuity formulae [5] as applied to physical region singularities. So I shall write

IV Spectral assumption $\Leftrightarrow$ Cutkosky formulae .

The current point of view seems to be that I and III constitute the most satisfactory choice of axioms since they can now be stated precisely and have a direct physical significance. From my point of view this is not satisfactory because it seems to me that the two statements I and II are not independent of each other. It follows that one must show that

i) I and III are consistent .

However, I think it would be even better to find a weaker version of I which could be valid in conjunction with III to prove I. This is why I have insisted on presenting my relatively vague assumption II because I had originally hoped to show something like [15]

ii) II + III ⇒ I .

It could be that in order to formulate II better, one should abstract the information contained in the linear program of axiomatic quantum field theory, but this has not yet been done completely.

Finally one would like to show

iii) I + III ⇒ IV .

Of course if (ii) and (iii) had been demonstrated, we would have

iv) II + III ⇒ I + IV .

This was what I originally wanted to do, and it would seem to be the ideal situation in that the maximum amount of information is deduced from the minimal reasonable assumptions. In fact a lot of work has been done to this end, to prove (iv), and later on I shall illustrate the arguments used, trying to show how the mathematical concepts and techniques developed relate to the recently developed hyperfunction theory. I shall also make the point that it is probably possible now to construct rigorous proofs of statement (i) and (ii) by combining the existing (physicists) arguments with the more refined and precise methods of hyperfunction theory.

3. Connectedness structure of the S-matrix and the unitarity equations

Pham has mentioned the S-matrix. A particular matrix element referring to given initial and final configurations of particles is the quantum mechanical probability amplitude [6,16] for the transition between these states. The conservation of probability, i.e. the fact that unity must be the total probability for some outcome to a scattering experiment, leads to the fact that S is unitary:

$$SS^{\dagger} = S^{\dagger}S = 1 \ .$$

In terms of S-matrix elements these equations are fairly complicated (and this will be important -- in fact where hyperfunction theory can help). The index summation implied in the matrix product involves a sum over the possible sets of particles in the intermediate states and also an integration over the momentum of each intermediate particle, consistent with it always being on its mass shell:

$$\frac{1}{(2\pi)^3}\int d^4p \ \theta(p^0)\delta(p^2 - m^2) \ .$$

Owing to energy and momentum conservation, this summation and integration is of finite range since it is restricted by the energy and momenta of the outside particles.

There are really even more terms in the unitarity equations than I have mentioned, since the S-matrix elements themselves break up into separate parts describing the different possibilities of subsets of particles colliding or missing each other altogether. This can best be represented by introducing a "bubble" notation, which is a sort of pictorial representation of the scattering process [15]. The S-matrix element is written

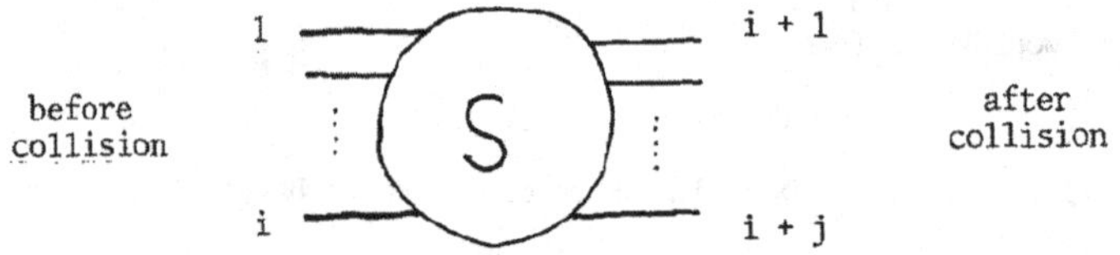

and the "cluster decomposition" just mentioned can be illustrated by

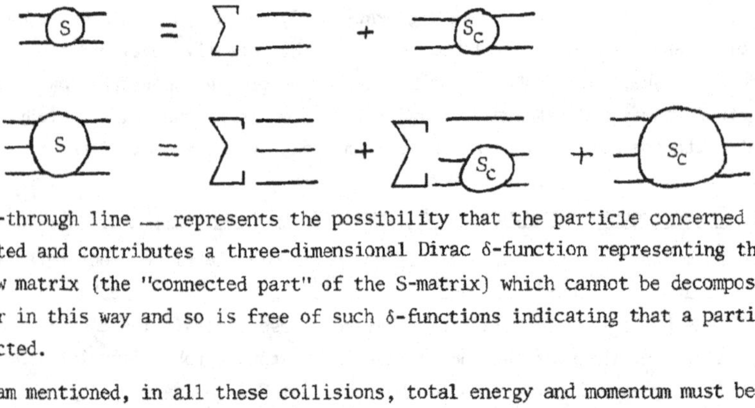

A straight-through line — represents the possibility that the particle concerned is not deflected and contributes a three-dimensional Dirac δ-function representing this. S_c is a new matrix (the "connected part" of the S-matrix) which cannot be decomposed any further in this way and so is free of such δ-functions indicating that a particle is undeflected.

As Pham mentioned, in all these collisions, total energy and momentum must be conserved, and this is represented by the fact that S_c contains as a factor just one four-dimensional δ-function guaranteeing over-all energy and momentum conservation:

$$\langle S_c \rangle = -i(2\pi)^4 \, \delta\left(\sum_{i \in I} p_i - \sum_{j \in J} p_j \right) \langle A \rangle \ .$$

The matrix elements of this new operator A are the fundamental quantities of the theory. They are the hyperfunctions defined on the mass shell manifold $M_{(IJ)}$ and at many physical points are in fact locally analytic.

Inserting these definitions of A into the unitarity equations, one finds the following relations: for example,

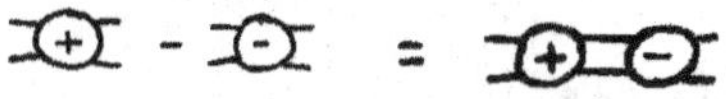 if $(2\,m)^2 < (p_1 + p_2)^2 < (3\,m)^2$

 if $(3\,m)^2 < (p_1 + p_2)^2 < (4\,m)^2$

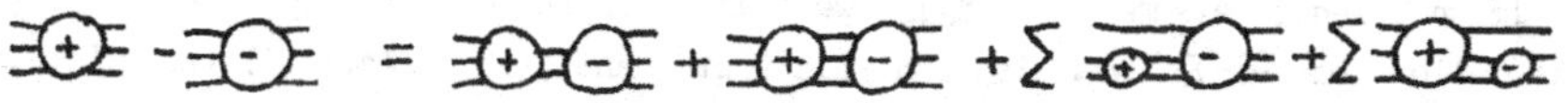

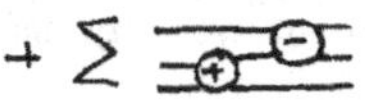 if $(3\,m)^2 < (p_1 + p_2 + p_3)^2 < (4\,m)^2$.

These are just two examples of the infinite system of non-linear integral equations satisfied by the hyperfunctions A(p) -- the examples we shall use later. In terms of the hyperfunctions A, the way to read off the terms of the above equations is given by the following rules [15]:

i) I  J $\to A(p_I, p_J)$

ii) I  J $\to A(p_J, p_I)^*$

iii) each internal line $\to -2\pi i\ \theta(p^0)\delta(p^2 - m^2)$

iv) each independent loop $\int \frac{id^4k}{(2\pi)^4}$

v) $\frac{1}{S_D}$ where S_D is the symmetry number of the diagram (n! for n identical particles) .

These equations, together with the ones we have not written down, have such a rich structure that it is perhaps not very surprising that we can deduce so much about the singular spectrum of the hyperfunctions A.

4. The pole singularity

This is the simplest example with which to illustrate the basic ideas.

Consider the causal configuration

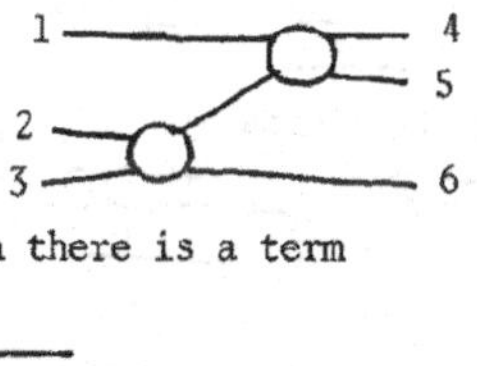

We notice that in the unitarity equation there is a term

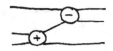

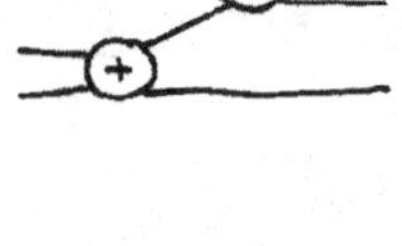

which by our rules is

$$A(p_2,p_3;\ p_2+p_3-p_6,p_6)(-2\pi i)\theta(p_2+p_3-p_6)\delta[(p_2+p_3-p_6)^2-m^2]$$
$$A(p_4,p_5;\ p_2+p_3-p_6,p_1)^* \ .$$

This is microanalytic except in the codirections u and -u associated with the above diagram, because of the δ-function.

If we tried to assume that  were analytic at $(p_2+p_3-p_6)^2 = m^2$, it would follow that  was also analytic there, and also in fact all the other terms of the equation except for the one above. This would then constitute a contradiction and illustrates our previous statement that microanalyticity and unitarity are not independent.

Now let us assume that  satisfies the microanalytic hypothesis I. Then at $(p_2+p_3-p_6)^2 = m^2$ it is microanalytic in all codirections except u. This is the same as saying in the old-fashioned language that it has a $+i\varepsilon$ prescription in the variable $(p_2+p_3-p_6)^2$. We can now examine all the other terms in the unitarity equation and find the following:

microanalytic in all directions except	u	-u	u	-u	u and -u	0
or in the old language	$+i\varepsilon$	$-i\varepsilon$	$+i\varepsilon$	$-i\varepsilon$	no $i\varepsilon$ prescription	analytic

So

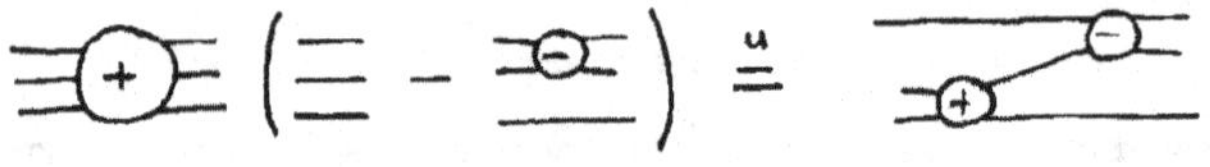

This means that the difference between the two sides of the equation is microanalytic in the u-direction (since the terms omitted are microanalytic in fact in all directions except the -u direction).

Now multiply on the right by

(this does not affect the microanalytic properties of the above terms) and use

to find

As Pham explained, this implies

the old-fashioned way of writing the result of the above argument, which dates from 1963 when stated in the old language as due to myself [15]. There is another later way of rearranging the argument due to Coster and Stapp [4]: First do the multiplication on the right and then examine the analytic properties.

5. The triangle singularity

This is slightly more complicated and illustrates some new features [11].

Consider the causal configuration

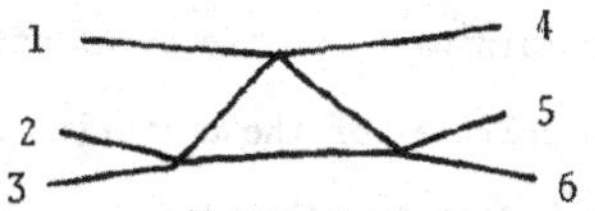

The corresponding values of p_I, p_J just satisfy an equation in the three variables $(p_2 + p_3)^2$, $(p_5 + p_6)^2$, and $(p_1 - p_4)^2$. Let us fix $(p_1 - p_4)^2$ at some physical (negative) value and look at the real values of the other two variables.

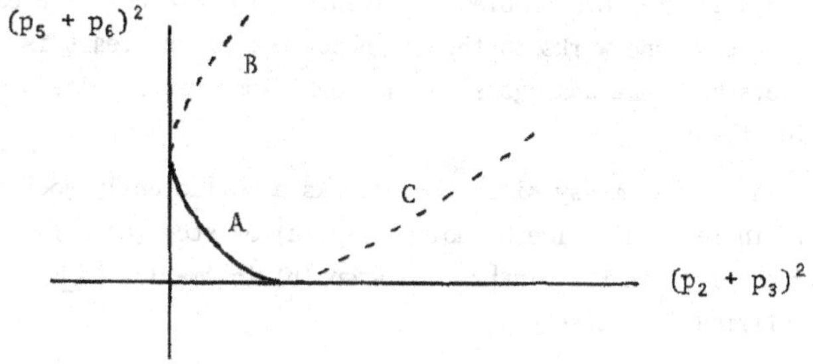

The causal (positive α) arc A is an arc of a hyperbola subtended between the lines $(p_2 + p_3)^2 = (2m)^2$ and $(p_5 + p_6)^2 = (2m)^2$ and should be singular. The other, non-causal arcs, B and C, have mixed α's, and should be non-singular (according to the microanalyticity postulate I).

When we look at the unitarity equation we find three terms each of which might be singular on any of the above arcs, A, B, and C; namely,

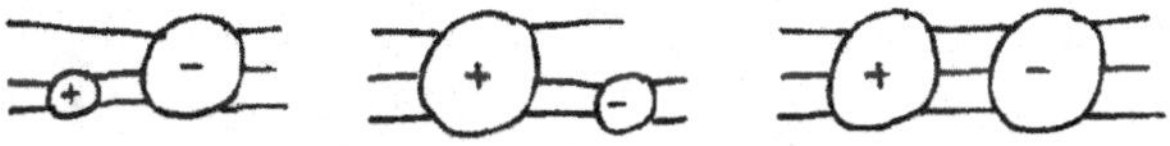

Then we analyse their analytic structure and make arguments analogous to those made for the pole singularity, and find

$$\text{disc} \quad \ldots \ldots = \begin{cases} \ldots \ldots & \text{on A} \\ 0 & \text{on B and C} \end{cases}$$

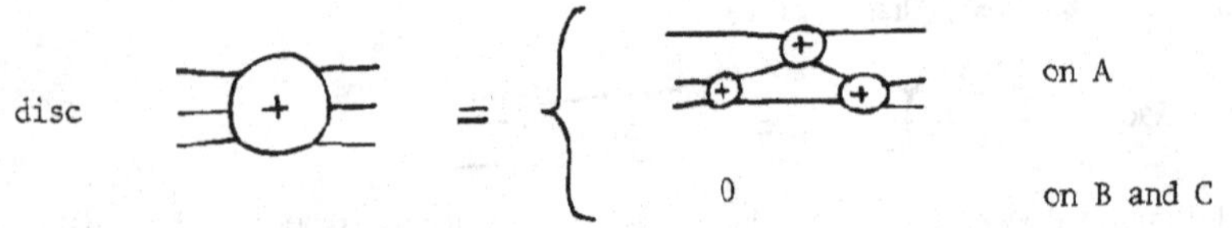

even if we make the weaker is assumption II. Thus all of statements (i)-(iv) can be checked in this case. More details are given in reference 11.

Thus we have seen how the various statements (i), (ii), and (iii) can be deduced in special cases, i.e. for specific S-matrix elements in specific parts of their physical region. These arguments have been considerably generalized [1,2], and there are always, as above, two main parts to the argument:

a)   the analysis of the analytic properties of the unitarity integrals;

b)   algebraic rearrangements using unitarity relations.

To achieve (a) one must know how to multiply hyperfunctions, integrate the result over a manifold (phase space), and study the microanalytic properties of the result. The physicists' solution to this problem is given in [1] together with earlier references. The fact that one works in the neighbourhood of the reals is continually exploited. I understand that analogous and no doubt improved results now emerge from hyperfunction theory.

Step (b) is in general messy since there lacks a sufficiently good notation for dealing with it. There is a choice of doing step (a) or step (b) first; step (a) was habitually done first in Britain [2] and step (b) in America [4]. The second step is then simplified (relatively).

To sum up, the unitarity of S-matrix theory constitutes a very interesting set of non-linear equations for the S-matrix. It seems that the resultant singular spectrum can be evaluated by techniques akin to hyperfunction theory, and that the more precise methods of this theory could be used to tidy up the arguments.

References

[1] M.J.W. Bloxham, D.I. Olive and J.C. Polkinghorne, S-matrix singularity structure in the physical region: I. Properties of multiple integrals, J. Math. Phys. 10 (1969) 494-502.

[2] M.J.W. Bloxham, D.I. Olive and J.C. Polkinghorne, S-matrix singularity structure in the physical region: II. Unitarity integrals, J. Math. Phys. 10 (1969) 545-552. III. General discussion of simple Landau singularities J. Math. Phys. 10 (1969) 553-561.

[3] G.F. Chew, S-matrix theory of strong interactions (W.A. Benjamin, New York, 1961).

[4] J. Coster and H.P. Stapp, Physical region discontinuity equations for many-particle scattering amplitudes: I. J. Math. Phys. 10 (1969) 371-396, II. J. Math. Phys. 11 (1970) 1441-1463.

[5] R.E. Cutkosky, Singularities and discontinuities of Feynman amplitudes, J. Math. Phys. 1 (1960) 429-433.

[6] R.J. Eden, P.V. Landshoff, D.I. Olive and J.C. Polkinghorne, The analytic S-matrix (Cambridge University Press, 1966).

[7] P. Goddard, Recent progress in the theory of dual resonance models, Proc. of the Aix-en-Provence Int. Conf. on Elementary Particles, 1973.

[8] J. Gunson, Unitarity and on-mass-shell analyticity as a basis for S-matrix theories, I. J. Math. Phys. 6 (1965) 827-844. II. J. Math. Phys. 6 (1965) 845-851.

[9] D. Iagolnitzer, Talk given in these proceedings. See also, D. Iagolnitzer, Introduction to S-matrix theory, ADT, 1973.

[10] D. Iagolnitzer and H.P. Stapp, Macroscopic causality and physical region analyticity in S-matrix theory, Comm. Math. Phys. 14 (1969) 15-55.

[11] P.V. Landshoff and D.I. Olive, Extraction of singularities from the S-matrix, J. Math. Phys. 7 (1966) 1464-1477.

[12] F. Pham, Introduction à la microanalyticité de la matrice S, Talk given in these proceedings.

[13] F. Pham, Singularités des processus de diffusion multiple, Ann. Inst. Henri Poincaré 6A (1967) 89-204.

[14] J.C. Polkinghorne, Analyticity and unitarity, Nuovo Cimento 23 (1962) 360-367; II. Nuovo Cimento 25 (1962) 901-911.

[15] D. Olive, Exploration of S-matrix theory, Phys. Rev. 135 B (1964) 745-760.

[16] D. Olive, Singularities in relativistic quantum mechanics, Proceedings of Liverpool Singularities Symposium II, Lecture Notes in Mathematics 209 (Springer Verlag, 1971), 244-269.

[17] H.P. Stapp, Derivation of the CPT theorem and the connection between spin and statistics from postulates of the S-matrix theory, Phys. Rev. 125 (1962) 2139-2162.

[18] J. Schwarz, Dual resonance theory, Physics Reports 8 C (1973) 269-335.

GEOMETRY OF THE N POINT P SPACE FUNCTION OF QUANTUM FIELD THEORY

H. Epstein
I.H.E.S. - Bures sur Yvette

V. Glaser
CERN - Geneva

R. Stora
CERN - Geneva
and
C.N.R.S. - Marseille

The geometry of the n point p space function of quantum field theory was first simultaneously developed by Araki [1] and Ruelle [2], following previous but less extensive work by Polkinghorne [3] and Steinman [4]. The construction of these authors which can be found in original papers and reviews is intuitively based on a preliminary study of the one-dimensional space-time situation - conventional non-relativistic quantum mechanics - which leads to the relativistic situation studied by these authors.

A more recent approach [5] is suggested by perturbation theory studies [6] as well as experience gained in the study of local analyticity properties of the scattering amplitudes in quantum field theory [7]. One natural way to supplement the Wightman axioms [8] - or, with minor modifications which do not affect analyticity statements, to exploit the Haag-Araki [9] axioms - is to assume the existence of operator valued distributions depending on n space-time points $(x_1 \ldots x_n) = X$, conventionally called time-ordered products of n field operators, whose products leave stable a dense domain in Hilbert space, which contains the vacuum vector Ω. Time-ordered products are assumed to fulfil the causal factorization property [6]:

$$T(X) = T(I)\,T(I')$$

if $I \gtrsim I'$ where $I' = X \setminus I$ and $I \gtrsim I' = (x \mid x_i - x_j \in \bar{V}^- \ \forall i \in I,\ j \in I')$. It is the latter condition that undergoes a minor modification in the Haag-Araki formulation of Q.F.T. One can thus naturally introduce the partially retarded operator [7]

$$R_I(X) = T(X) - T(I')\,T(I)$$

with support $(x \mid x_i - x_j \in \bar{V}^+$ some $i \in I,\ j \in I')$. Let now

$$(\Omega, R_I(X)\Omega) = r_I(X)\,;\quad (\Omega, T(X)\Omega) = t(X)$$

Applying spectrum properties to Fourier transforms, one obtains

$$\tilde{r}_I(P) = \tilde{t}(P) \quad \text{if} \quad p_I \notin S_I^-$$

where $S_I^- = \{0\} \cup \bar{V}_I^-$, $\bar{V}_I^-$ being the closed convex hull of the lower sheet of a hyperboloid of mass M_I depending on selection rules and spectrum.

These properties can be extended to more general expressions which we shall now construct.

1. - THE ALGEBRA GENERATED BY TIME-ORDERED PRODUCTS

One considers the vector space $\mathcal{T}$ spanned by ordered monomials

$$\prod_{k=1}^{k=\nu} T(I_k) \equiv T(I_1)\dots T(I_\nu)\,,\quad I_1 \cup \dots \cup I_\nu = X\,,\quad I_i \cap I_j = \phi\,,\quad i,j = 1\dots\nu\,,$$

whose generic element will be denoted Θ. If $I \subseteq X$, one defines $\hat{I}$ by linear extension of

$$\hat{I} \prod_{k=1}^{k=\nu} T(I_k) = \prod_{k=1}^{k=\nu} T(I \cap I_k)\, T(I' \cap I_k)$$

with the convention $\hat{X} = \hat{\emptyset} = 1$, $T(\emptyset) = 1$. One has the following properties :

(i) $\hat{I}^2 = \hat{I}$

(ii) $\hat{I}_1 \hat{I}_2 = \hat{I}_2 \hat{I}_1 \Longleftrightarrow I_1 \subseteq I_2$ or $I_2 \subseteq I_1$

$$\left(I_1' \cap I_2 = I_1 \cap I_2' \rightarrow I_1' \cap I_2 = I_1 \cap I_2' \cap I_2 = \phi\right)$$

(iii) $$\left(1 - \widehat{I_1 \cup I_2}\right) \prod_{r=1}^{r=\nu} \hat{K}_r\, \hat{I}_1\, \hat{I}_2 = \prod_{r=1}^{r=\nu} \hat{K}_r\, \hat{I}_1\, \hat{I}_2 \left(1 - \widehat{I_1 \cup I_2}\right)$$

$$= \prod_{r=1}^{r=\nu} \hat{K}_r\, \hat{I}_1 \left(1 - \widehat{I_1 \cup I_2}\right) \hat{I}_2 = 0$$

(iv) same as (iii) with $\cup \rightarrow \cap$

(v) $\hat{I}'\hat{I} = \hat{I}$

$\hat{I}\,\hat{I}' = \hat{I}'$

(vi) T(X) is cyclic :

$$T(K_1)\dots T(K_\nu) = \hat{I}_1 \dots \hat{I}_\nu \; T(X)$$

$$I_r = \bigcup_{s=1}^{s=r} K_s$$

This allows to define a partial order on $\mathcal{T}$:

$$T(K_1)\dots T(K_\nu) > T(L_1)\dots T(L_\mu)$$

if the partition $(L_1 \dots L_\mu)$ refines the partition $(K_1 \dots K_\nu)$.

(vii) If $I_1 \cap I_2 = \phi$, $\hat{I}_1 \hat{I}_2 = \widehat{I_1 \cup I_2}\, \hat{I}_2 = \hat{I}_2\, \widehat{I_1 \cup I_2}$

(viii) If $I_1 \cup I_2 = X$, $\hat{I}_1 \hat{I}_2 = \widehat{I_1 \cap I_2}\, \hat{I}_2 = \hat{I}_2\, \widehat{I_1 \cap I_2}$

(ix) Eigenstates of $\hat{I}$:

$$\hat{I}\,\Theta = \Theta \overset{\text{Def.}}{\Longleftrightarrow} \Theta \in \mathcal{N}(I).$$

vectors of the form

$$\Theta = \prod_{k=1}^{k=\nu} T(I_k)$$

with $I_k \subset I$ or $I_k \subset I'$ belong to $\mathcal{N}(I)$.

Conversely, all elements of $\mathcal{N}(I)$ are of this form :

If $\hat{I} \prod_{k=1}^{k=\nu} T(I_k) \neq \prod_{k=1}^{k=\nu} T(I_k) \in \mathcal{D}(I)$,

$$(1-\hat{I}) \prod_{k=1}^{k=\nu} T(I_k) \neq 0, \quad \hat{I}(1-\hat{I}) \prod_{k=1}^{k=\nu} T(I_k) = 0.$$

In other words, either a monomial is non-decomposable [in $\mathcal{N}(I)$], and is an eigenstate of $\hat{I}$ with eigenvalue 1, or it is decomposable [in $\mathcal{D}(I)$] and there corresponds to it an eigenstate of $\hat{I}$ with eigenvalue 0 - such eigenstates of $\hat{I}$ are thus put in a one-to-one correspondence with all partitions of X and thus span the whole of $\mathcal{C}$.

2. - PRECELLS [5),10)]

We now look for generalizations of the r_I's, with support properties in x space, and with coincidence properties in p space with $\tilde{t}(P)$.

Def [5)]

A precell $\mathcal{S}$ is a set of proper parts of X ($\neq \emptyset$, X) such that

1) If $I \in \mathcal{S}$, $I' \notin \mathcal{S}$.
Let $\mathcal{S}' = \{ I' \mid I \in \mathcal{S} \}$.
If $I \notin \mathcal{S}$, then $I' \in \mathcal{S}$.

2) If $I_1 \in \mathcal{S}$, $I_2 \in \mathcal{S}$, $I_1 \cap I_2 = \emptyset$
then $I_1 \cup I_2 \in \mathcal{S}$.

3) If $I_1 \in \mathcal{S}$, $I_2 \in \mathcal{S}$, $I_1 \cup I_2 = X$
then $I_1 \cap I_2 \in \mathcal{S}$

[a consequence of 1) and 2) applied to $\mathcal{S}'$]

Def

$$\Theta_{\mathcal{S}'} = \prod_{I \in \mathcal{S}} (1 - \hat{I}) \, \Theta$$

$$\theta_{\mathcal{S}'} = (\Omega, \Theta_{\mathcal{S}'} \Omega)$$

Precell Lemma [5)] (elementary)

If $I_1 \in \mathcal{S},\ I_2 \in \mathcal{S}$

Then $I_1 \cap I_2 \in \mathcal{S}$ or $I_1 \cup I_2 \in \mathcal{S}$

Def (paracell) [10)]

A paracell is a set $\mathcal{S}$ of proper parts of X for which the precell lemma holds.

Th

If $\mathcal{S}$ is a paracell, the product

$$\prod_{I \in \mathcal{S}} (1 - \hat{I})$$

is independent of the order of the factors.

Sketch of proof

Let $I_1 \ldots I_N$ be an arbitrary ordering of $\mathcal{S}$, we have to show that for any permutation π

$$(1-\hat{I}_1)\ldots(1-\hat{I}_N) = (1-\hat{I}_{\pi(1)})\ldots(1-\hat{I}_{\pi(N)})$$

Suppose π is the transposition $N \rightleftarrows N-1$. The property is true by (ii), (iii), (iv). Suppose the property has been proved for all π 's such that

$$\pi(1) = 1\ ,\ \pi(2) = 2, \ldots\ \pi(N-p) = N-p,$$

The passage from p to p+1 follows easily from the induction hypothesis and application if (ii), (iii), (iv). In fact the proof shows the following.

Lemma

Let $A_1 \ldots A_N$ be a sequence of not necessarily distinct proper subsets of X, such that for every pair i, j, there exists k such that $A_k = A_i \cup A_j$ or $A_k = A_i \cap A_j$, then for every permutation π of $1 \ldots N$

$$(1-\hat{A}_1)\ldots(1-\hat{A}_N) = (1-\hat{A}_{\pi(1)})\ldots(1-\hat{A}_{\pi(N)}) \ .$$

- Support properties

Lemma

$$(\mathbb{1}-\hat{I})\,\Theta_{\mathcal{S}'}=0 \quad \text{if} \quad I'\gtrsim I \;,\; I\in\mathcal{S}$$

hence support $\Theta_{\mathcal{S}'}\subset C_I=\{x\,/\,x_i-x_j\in\overline{V}_+ \text{ , some } i\in I \text{ , } j\in I'\}$

$$=\bigcup_{\substack{i\in I\\ j\in I'}}(x\,/\,x_i-x_j\in\overline{V}^+)=\bigcup_{\substack{i\in I,\\ j\in I'}}C_{ij}$$

(use causal factorization property).

- Corollary

$$\Theta_{\mathcal{S}'}=\prod_{I\in\mathcal{S}}(\mathbb{1}-\hat{I})\,\Theta$$

has support contained in

$$\bigcap_{\substack{I\in\mathcal{S}\\ I'\in\mathcal{S}'}}\;\bigcup_{\substack{i\in I\\ j\in I'}}C_{ij}$$

This support can be analyzed as follows. A choice [5),7)] h is a map $I\to X$ such that $h(I)\in I$. Let

$$C_h^{\mathcal{S}}=\bigcap_{I\in\mathcal{S}}C_{h(I)\,h(I')}$$

With this notation

$$\bigcap_{\substack{I\in\mathcal{S}\\ I'\in\mathcal{S}'}}\;\bigcup_{\substack{i\in I\\ j\in I'}}C_{ij}=\bigcup_h C_h^{\mathcal{S}}$$

Choices h which yield the support can be restricted to choices compatible [5] with $\mathcal{S}$ as follows. Let

$$\mathcal{S}^+ = \{ j \in X, \{j\} \in \mathcal{S} \}$$

$$\mathcal{S}^- = X \setminus \mathcal{S}^+ = \mathcal{S}'^+$$

One can show [5] that for each choice h, there exists a "compatible choice" h' such that

$$h'(I) \in \mathcal{S}^+ \quad \text{if } I \in \mathcal{S}$$

$$h'(I) \in \mathcal{S}^- \quad \text{if } I \in \mathcal{S}'$$

and

$$C_h^{\mathcal{S}} \subset C_{h'}^{\mathcal{S}}$$

from which there follows

$$\text{supp.}\, \Theta_{\mathcal{S}'} \subset \bigcup_{h \text{ compatible}} C_h^{\mathcal{S}}$$

One can furthermore show [5] that each $C_h^{\mathcal{S}}$ is of the form

$$\bigcap_{\substack{i \in \mathcal{S}^+ \\ j \in \mathcal{S}^- \\ (ij) \in \mathcal{T}_h}} \{ x \mid x_i - x_j \in \bar{V}^+ \}$$

where $\mathcal{T}_h$ is a tree graph with vertices $i \in X$ and links (ij), $i \in \mathcal{S}^+$, $j \in \mathcal{S}^-$. Such trees were first discovered by Bros [11].

- Analyticity properties

It is easy to see that the Fourier transform of a translation invariant distribution with support in C_h can be extended into a holomorphic function defined on the hyperplane

$$p_X \equiv \sum_{i=1}^{n} p_i = 0$$

and analytic in the tube with imaginary basis $\tilde{C}_h$ dual to C_h, given by the parametric form [1,9]

$$p = \sum_{(ij) \in \mathcal{T}_h} S_{(ij)} \, t_{(ij)}$$

where $$\mu_{(ij)} \in V^+, \quad S_{(ij)} = \left(S^1_{(ij)}, \dots S^n_{(ij)} \right),$$
$$S^k_{(ij)} = \delta^k_i - \delta^k_j$$

This parametric form is due to Araki [1),9)] and has the following nice property : if a translation invariant distribution has support $C^{\mathcal{S}}_h \cap C^{\mathcal{S}}_{h'}$, its Fourier transform can be extended into the tube with imaginary basis

$$p = \sum S_\lambda \mu_\lambda$$

where $\mu_\lambda \in V^+$ and S_λ are the co-ordinates of exposed one-dimensional facets of the convex hull

$$S = t \sum_{(ij) \in C_h} S_{(ij)} \rho_{(ij)} + (1-t) \sum_{(ij) \in C_{h'}} S'_{(ij)} \rho'_{(ij)}$$

$$0 \leq t \leq 1, \quad 0 \leq \rho_{(ij)}, \quad 0 \leq \rho'_{(ij)}.$$

Alternatively, the tube corresponding to C_h can be described by the set of $n-1$ conditions of the type $\mathrm{Im}\, p_{I(ij)} \in V^+$, $I_{(ij)} \in \mathcal{S}$, obtained by writing

$$\sum_{i=1}^{i=n} p_i x_i = \sum_{(ij) \in \mathcal{T}_h} p_{I(ij)} (x_i - x_j) \mod \sum_{i=1}^{i=n} p_i .$$

A description of the latter kind does not exist for convex hulls, in general.

3. - SPECTRUM PROPERTIES OF $T_{\mathcal{S}}$

We look at Fourier transforms of vacuum expectation values, $\tilde{t}_{\mathcal{S}}(P)$ obtained for

$$\Theta = T(X) \;.$$

Expanding the product

$$T_{\mathcal{S}}(X) = \prod_{I \in \mathcal{S}'} (\mathbb{1} - \hat{I})\, T(X)$$

and looking at the last factor of each term, we see that

$$\tilde{t}_{\mathcal{S}}(P) = \tilde{t}(P) \qquad \text{if } p_I \notin S_I^- \quad \forall I \in \mathcal{S}$$

4. - DISCONTINUITY FORMULAE

Def

If $\mathcal{S}$ is a precell, I is called a boundary of $\mathcal{S}$ ($I \in \partial\mathcal{S}$) if

$$(\mathcal{S} \setminus \{I\}) \cup \{I'\}$$

is a precell.

Boundary Lemma [5)]

I is a boundary of $\mathcal{S}$ if

(i) $I_1 \cup I_2 = I \;,\; I_1 \cap I_2 = \emptyset \quad I_1 \neq I,\; I_1 \in \mathcal{S}$
$\rightarrow I_2 \in \mathcal{S}'$

(ii) $I_1' \cup I_2' = I',\; I_1' \cap I_2' = \emptyset \;,\; I_1' \neq I' \;,\; I_1' \in \mathcal{S}'$
$\rightarrow I_2' \in \mathcal{S}$

(i) and (ii) are equivalent to (i) and :

(iii) $$I_1 \cap I_2 = I\ ,\ I_1 \cup I_2 = X,\ I_1 \neq I\ ,\ I_1 \in \mathcal{S}$$
$$\rightarrow I_2 \in \mathcal{S}'$$

If I is a boundary of $\mathcal{S}$ we shall define

$$\mathcal{S}_I^+ = \mathcal{S}\ ,\quad \mathcal{S}_I^- = (\mathcal{S} \setminus \{I\}) \cup \{I'\}\ .$$
$$\mathcal{S}_I = (\mathcal{S} \setminus \{I\})$$

Using the commutativity of $\prod_{I \in \mathcal{S}}$ $(1-\hat{I})$ and leaving I as the last factor, one easily gets

$$\prod_{J \in \mathcal{S}_I^+} (1-\hat{J}) - \prod_{J \in \mathcal{S}_I^-} (1-\hat{J}) = \prod_{J \in \mathcal{S}_I} (1-\hat{J}) [\hat{I}' - \hat{I}]$$

Hence the commutator formula

$$T_{\mathcal{S}_I^{+}} - T_{\mathcal{S}_I^{-}} = \prod_{J \in \mathcal{S}_I} (1-\hat{J}) [T(I'), T(I)]\ .$$

Write now

$$\hat{J} = \hat{J}_{I'} \cdot \hat{J}_I = \widehat{J \cap I'} \cdot \widehat{J \cap I}$$

as a direct product which allows to write, if $K \subset I$, $K' \subset I'$

$$\hat{J}\, T(K)\, T(K') = \hat{J}_{I'} \cdot \hat{J}_I\, T(K)\, T(K') = \hat{J}_I\, T(K) \cdot \hat{J}_{I'}\, T(K')$$

where

$$\hat{J}_I\, T(K) = T(K \cap J \cap I)\, T(K \cap J' \cap I)$$

$$\hat{J}_{I'}\, T(K') = T(K' \cap J \cap I')\, T(K' \cap J' \cap I')$$

Let now

$$\Sigma_I = \{ K \mid K \subset I ,\ K \in \mathcal{S}_I \}$$

$$\Sigma_{I'} = \{ K \mid K \subset I' \ \ K \in \mathcal{S}_I \}$$

These are precells from the precell property of $\mathcal{S}$ and the boundary propert of I.

Now if $J \in \mathcal{S}_I$, either $J \cap I$ is in Σ_I or $J \cap I'$ is in $\Sigma_{I'}$ (or both) because if it were not so, both $J \cap I$ and $J \cap I'$ would be in $\mathcal{S}'$ and since they are disjoint, their union, J, would be in $\mathcal{S}'$ which is not true. Hence J is of one of the following types :

(i) $K \cup K'$, $K \in \Sigma_I$, $K' \in \Sigma_{I'}$.

(ii) $K \cup \emptyset$, $K \in \Sigma_I$.

(iii) $\emptyset \cup K'$, $K' \in \Sigma_{I'}$.

(iv) $K \cup I'$, $K \in \Sigma_I$.

(v) $I \cup K'$, $K' \in \Sigma_{I'}$.

(vi) $K \cup L'$, $K \in \Sigma_I$, $L' \subset I'$, $(L' \neq I', \emptyset)$.

(vii) $L \cup K'$, $L \subset I$, $(L \neq I, \emptyset)$, $K' \in \Sigma_{I'}$.

Using again the precell property of $\mathcal{S}$, the commutativity of $\prod_{J \in \mathcal{S}_I} (1-\hat{J})$ and the idempotency identities

$$(1 - \hat{K})^2 = (1 - \hat{K}) \ , \quad (1 - \hat{K}')^2 = 1 - \hat{K}'$$

we can write

$$\prod_{J \in \mathcal{S}_I} (1 - \hat{J}) = \prod_{\substack{K \in \Sigma_I \\ K' \in \Sigma_{I'}}} (1 - \hat{K}' \hat{K})(1 - \hat{K})(1 - \hat{K}') \ldots$$

$$\ldots \quad \cdot \prod_{\substack{K \in \Sigma_{I'} \\ L \subseteq I}} (1 - \hat{K}' \cdot \hat{L})(1 - \hat{K}') \cdot \prod_{\substack{L' \subset I' \\ K \in \Sigma_I}} (1 - \hat{L}' \cdot \hat{K})(1 - \hat{K})$$

$$= \prod_{\substack{K \in \Sigma_I \\ K' \in \Sigma_{I'}}} (1 - \hat{K}')(1 - \hat{K})$$

where we have used

$$(1 - \hat{K}' \hat{K})(1 - \hat{K})(1 - \hat{K}') = (1 - \hat{K}') \cdot (1 - \hat{K})$$

$$(1 - \hat{K}' \cdot \hat{L})(1 - \hat{K}') = (1 - \hat{K}')$$

$$(1 - \hat{L}' \cdot \hat{K})(1 - \hat{K}) = (1 - \hat{K})$$

on account of the idempotency

$$\hat{K}^2 = \hat{K} \quad , \quad \hat{K}'^2 = \hat{K}'$$

The commutativity of the factors $(1-\hat{K})$, $(1-\hat{K}')$ allows to split these factors :

$$T_{\mathcal{S}_I^{+'}} - T_{\mathcal{S}_I^{-'}} = \prod_{\substack{K \in \Sigma_I \\ K' \in \Sigma_{I'}}} (1 - \hat{K})(1 - \hat{K}') \left[T(I'), T(I) \right]$$

$$= \left[\prod_{K' \in \Sigma_{I'}} (1 - \hat{K}') T(I') , \prod_{K \in \Sigma_I} (1 - \hat{K}) T(I) \right]$$

$$= \left[T_{\Sigma'_{I'}}(I') \quad , \quad T_{\Sigma'_I}(I) \right]$$

This is the so-called Ruelle discontinuity formula [2),5)].

5. - STEINMANN's IDENTITIES [4),1),2)]

Let I, J such that $I \not\subset J$, $I \not\subset J'$, $J \not\subset I$, $J \not\subset I$, hence $I' \not\subset J$, $I' \not\subset J'$, $J \not\subset I'$, $J' \not\subset I'$. Consider four cells admitting I, J, as boundaries :

$$\mathcal{S}_{IJ}^{++} = \mathcal{S}_{IJ} \cup I \cup J = \mathcal{S}_I \cup J = \mathcal{S}_J \cup I$$

$$\mathcal{S}_{IJ}^{+-} = \mathcal{S}_{IJ} \cup I \cup J' = \mathcal{S}_I \cup J' = \mathcal{S}_{J'} \cup I$$

$$\mathcal{S}_{IJ}^{-+} = \mathcal{S}_{IJ} \cup I' \cup J = \mathcal{S}_{I'} \cup J = \mathcal{S}_J \cup I'$$

$$\mathcal{S}_{IJ}^{--} = \mathcal{S}_{IJ} \cup I' \cup J' = \mathcal{S}_{I'} \cup J' = \mathcal{S}_{J'} \cup I'$$

Then

$$T_{\mathcal{S}_{IJ}^{++'}} - T_{\mathcal{S}_{IJ}^{-+'}} = \left[T_{\Sigma_I^{+'}}(I),\ T_{\Sigma_{I'}^{+'}}(I') \right]$$

where

$$\Sigma_I^+ = \{ K \mid K \subset I ,\ K \in \mathcal{S}_{IJ} \}$$
since $J \not\subset I$

$$\Sigma_{I'}^+ = \{ K \mid K \subset I' ,\ K \in \mathcal{S}_{IJ} \}$$
since $J \not\subset I'$

Similarly

$$T_{\mathcal{S}_{IJ}^{+-'}} - T_{\mathcal{S}_{IJ}^{--'}} = \left[T_{\Sigma_I^{-'}}(I),\ T_{\Sigma_{I'}^{-'}}(I') \right]$$

where

$$\Sigma_I^- = \{ K \mid K \subset I ,\ K \in \mathcal{S}_{IJ} \}$$
since $J' \not\subset I$

$$\Sigma_{I'}^- = \{ K \mid K \subset I' ,\ K \in \mathcal{S}_{IJ} \}$$
since $J' \not\subset I'$

Hence

$$\Sigma_I^+ = \Sigma_I^-$$

$$\Sigma_{I'}^+ = \Sigma_{I'}^-$$

and

$$T_{\mathcal{S}_{IJ}^{++}} - T_{\mathcal{S}_{IJ}^{+-}} - T_{\mathcal{S}_{IJ}^{-+}} + T_{\mathcal{S}_{IJ}^{--}} = 0$$

The so-called Steinmann-Ruelle relations [1),2)].

6. - CELLS [1),2)]

The interpretation of $\tilde{t}(P)$ as a piecewise boundary value of analytic functions goes via the consideration of cells.

Def [5)]

A precell $\mathcal{S}$ is a cell if, in R^n, on the hyperplane $S_X = \sum_i^n S_i = 0$, the set of conditions

$$S_I = \sum_{i\in I} S_i > 0 \quad \forall\, I \in \mathcal{S}$$

is non empty. This set will be called $\Gamma_{\mathcal{S}}$. Clearly, if $S \in \Gamma_{\mathcal{S}}$ $S_I < 0$ if $I \in \mathcal{S}'$.

One can easily see that Steinmann-Ruelle relations can only connect together cells whose Γ's are within a hypercell Γ_I : $S_i > 0$ $i \in I$, $S_i < 0$ $i \in I'$, i.e., which all contain i, $i \in I$ and $[i]'$, $i \in I'$. The one dimensional boundaries of Γ_I are the vectors S_{ij}, $i \in I$, $j \in I'$, $S_{ij}^k = \delta_i^k - \delta_j^k$, we already met. One can prove that for a cell $\mathcal{S}$, the admissible $\tilde{C}_h$ are such that Γ_h :

$$S = \sum_{(ij)\in \mathcal{C}_h} S_{(ij)}\,\rho_{(ij)} \quad , \quad \rho_{(ij)} > 0$$

contain $\Gamma_{\mathcal{S}}$.

A more detailed result (Bros [11]) shows that for all $\mathcal{J}$'s such that $\Gamma_{\mathcal{J}} \subset \Gamma_I$ there is a decomposition of the form

$$\tilde{t}_{\mathcal{J}} = \sum_h \Theta_{\mathcal{J},h}\, \tilde{f}_h$$

where the $\tilde{f}_h$ are holomorphic in the tube with imaginary basis $\tilde{C}_h$ (the dual of C_h), where

$$\Theta_{\mathcal{J},h} = 1 \quad \text{if} \quad \Gamma_h \supset \Gamma_{\mathcal{J}}$$

$$\Theta_{\mathcal{J},h} = 0 \quad \text{if} \quad \Gamma_h \not\supset \Gamma_{\mathcal{J}}$$

The foregoing results are best summarized within the framework of the following geometrical construction.

Consider an n dimensional space $\{ S = (S_1, \dots, S_n) \}$ and the hyperplane

$$\sum_{i=1}^{i=n} S_i = 0$$

One draws all the hyperplanes

$$S_I \equiv \sum_{i \in I} S_i = 0 \quad (= S_{I'})$$

which bound geometrical cells $\Gamma_{\mathcal{J}}$ within which all S_I's have a prescribed sign. One can easily read off this diagram the various relevant S_{ij}'s in the preceding formulae.

Examples are shown on Figs. 1, 2, 3.

7. - TRUNCATION

One has still to make a slight modification in order to make the foregoing construction useful. The regions in p space where the $\tilde{t}$ (P) coincide with $\tilde{t}(P)$ do not cover the whole of P space because of the occurrence of the vacuum contribution $\{0\}$ to the spectrum S_I^+. One can however, define connected distributions [6] $\tilde{t}_c(P)$, $\tilde{t}_{\mathcal{S},c}(P)$ [which turn out to be identical to $\tilde{t}_{\mathcal{S}}(P)$] such that all previously established properties hold. The regions where $\tilde{t}_c(P)$ coincide with the $\tilde{t}_{\mathcal{S}}(P)$'s cover all of P space when $\mathcal{S}$ goes over all possible cells, S_I^+ being then reduced to $\bar{V}_I^+$. Although this is quite important, we shall not give any more detail here.

The bridge can be made with the construction of Haraki and Ruelle by establishing the following identity [5),12)]

$$T_{\mathcal{S}} = \sum_{\nu} (-)^{\nu+1} \sum_{\substack{I_1 \cup .. \cup I_\nu = X \\ I_j \cap I_k = \phi \\ K_r = I_1 \cup \dots \cup I_r \in \mathcal{S} \\ r = 1, .., \nu-1}} T(I_1) \dots T(I_\nu)$$

from which one easily recovers the spectrum properties, the discontinuity formula and the Steinmann identities, but from which support properties are hard to get.

REFERENCES

1) H. Araki - J.Math.Phys. 2, 163 (1961).

2) D. Ruelle - Nuovo Cimento 19, 356 (1961).

3) J.C. Polkinghorne - Nuovo Cimento 4, 216 (1956).

4) O. Steinmann - Helv.Phys.Acta 33, 257 (1960) ; ibid 33, 347 (1960).

5) J. Bros, H. Epstein, V. Glaser and R. Stora - to be published.

6) H. Epstein and V. Glaser - CERN Preprint TH. 1400 (1971).

7) J. Bros, H. Epstein and V. Glaser - Helv.Phys.Acta 45, 149 (1972).

8) R. Jost - "The General Theory of Quantized Fields", A.M.S. Providence (1965).

9) H. Araki - ETH Lectures, Zürich, unpublished.

10) Reference 5) introduces the more general notion of paracells which is a special case of the notion of cycle introduced by D. Ruelle, Ref. 2).

11) J. Bros - Thesis, Paris (1970) ; Lectures at RCP 25, Strasbourg, Mathematics Department, Vol. VIII (1969).

12) The first step in deriving this formula from the definition of Refs. 1), 2), was taken by C. Itzykson (1963), unpublished.

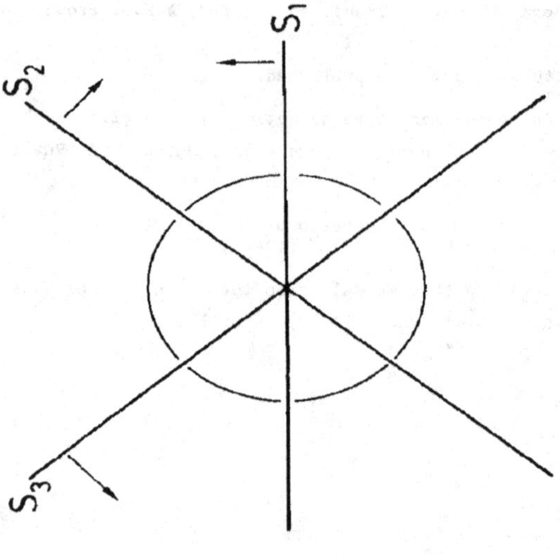

FIG.1
$n = 3$

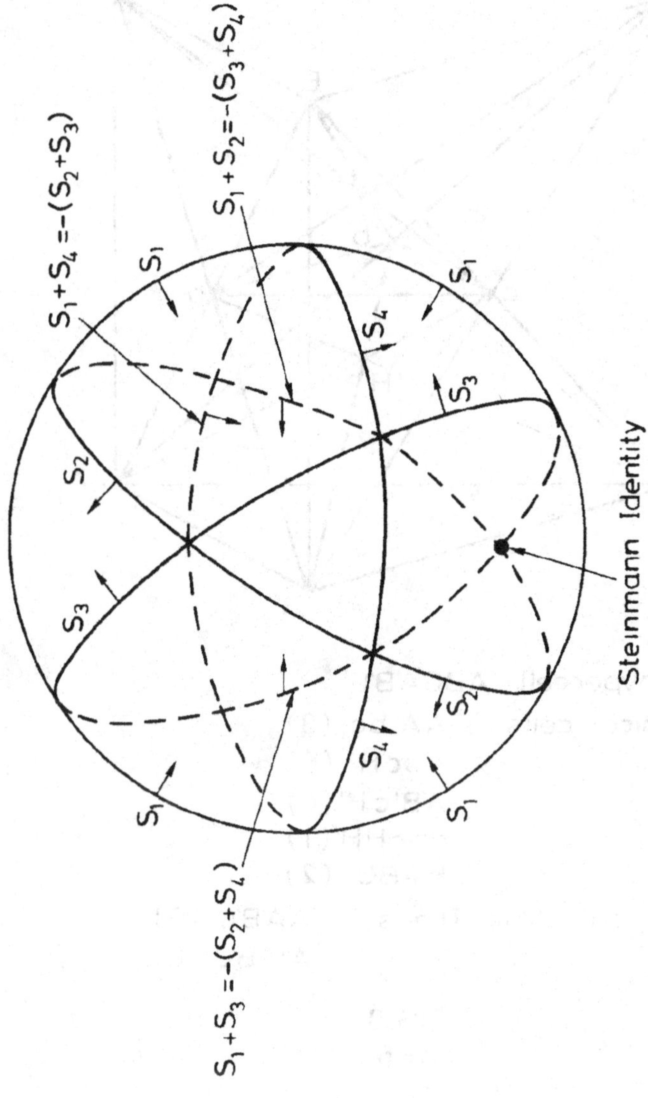
S₁ + S₄ = -(S₂+S₃)
S₁ + S₂ = -(S₃+S₄)
S₁ + S₃ = -(S₂+S₄)
S₁
S₁
S₁
S₁
S₂
S₂
S₃
S₃
S₄
S₄
Steinmann Identity

FIG. 2
$n = 4$

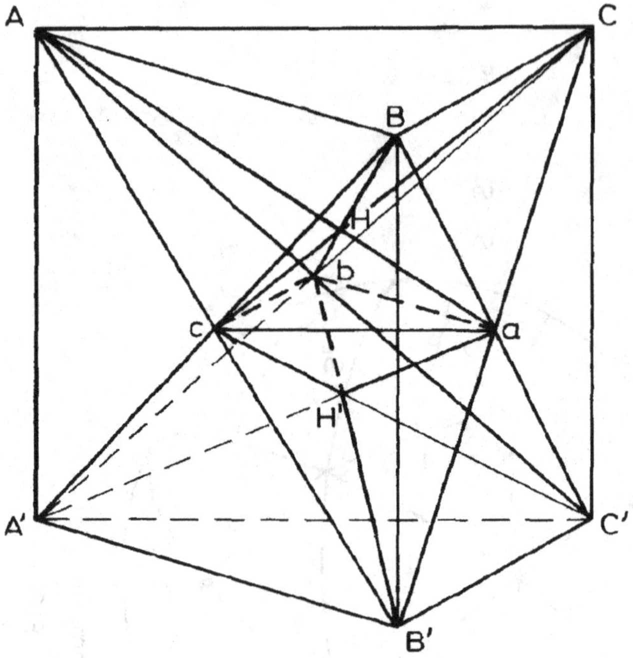

A hypercell ABCA'B'C'

Typical cells : AA'bc (3)

AbcH (6)

A'B'cH' (6)

abcHH' (1)

HABC (2)

Typical Bros Trees : AA'B'C' (6)

A'BB'C (6)

FIG. 3

$n=5$

SOME APPLICATIONS OF THE JOST-LEHMANN-DYSON THEOREM TO THE STUDY OF THE GLOBAL ANALYTIC STRUCTURE OF THE N POINT FUNCTION OF QUANTUM FIELD THEORY

R. Stora
CERN - Geneva
and
C.N.R.S. - Marseille

INTRODUCTION

Whereas the theory of hyperfunctions and microfunctions [1] is able to produce local information on the problem posed by the structure of the p space n point function of quantum field theory [2], it is quite obvious that the problem physicists are confronted with is of a global nature.

Most global results that are known at present can be derived from a few ingredients :

- the tube theorem,
- the semi-tube theorem,
- the edge-of-the wedge theorem,
- the vanishing of the first cohomology group of a Stein manifold,

several combinations of which are able to yield partial answers of a global nature. In some cases, more sophisticated applications of the continuity theorem were able to produce non-trivial analytic completions.

Standing aside is a beautiful piece of work due to Jost, Lehmann and Dyson (JLD) [3], which is by itself able to produce many of the results known by the previously mentioned methods and has, besides, a touch of originality and exotism which has made it almost impossible to generalize, excepting an unpublished work by Glaser. This can be taken as a proof that, if there is a general idea behind it, no one has been so far able to grasp it.

We shall furthermore see here some applications [4] of JLD's trick which have so far not been recovered by the more conventional methods.

In what follows, we shall first give a summary of JLD's work, then make the bridge between some of its consequences and results which can be obtained by analytic methods, and finally give some applications which, at present, cannot be treated otherwise.

1. - THE JOST-LEHMANN-DYSON (JLD) REPRESENTATION

A) - Let C be a temperate distribution in Minkowski configuration (x) space, with support the union of two closed opposite light cones

$$\bar{V}^{\pm} = \left\{ x \mid x^2 = (x^0)^2 - (x^1)^2 - (x^2)^2 - (x^3)^2 \equiv (x^0)^2 - \vec{x}^2 \geq 0,\ x^0 \gtrless 0 \right\}$$

The Fourier transform $\tilde{C}$ of C is the restriction to the hyperplane $\varkappa = 0$ of a tempered distribution Γ in five-dimensional momentum Minkowski space $(p, \varkappa)$, where $\varkappa$ is an additional space-like co-ordinate, even under reflections through the hyperplane $\varkappa = 0$, and solution of the wave equation

$$\square_{p,\varkappa}\, \Gamma = 0$$

where

$$\square_{p,\varkappa} = \frac{\partial^2}{(\partial p^0)^2} - \frac{\partial^2}{(\partial p^1)^2} - \frac{\partial^2}{(\partial p^2)^2} - \frac{\partial^2}{(\partial p^3)^2} - \frac{\partial^2}{(\partial \varkappa)^2}$$

Conversely, every solution Γ of the wave equation that is even in $\varkappa$ has a restriction to the hyperplane $\varkappa = 0$ whose Fourier transform has support $\bar{V}^+ \cup \bar{V}^-$. C's and Γ's with the stated properties are in a one-to-one correspondence

A*) - This situation can be generalized to the case where $\bar{V}^{\pm}$ is replaced by $\bar{V}^{\pm}_M = \left\{ x \mid x^2 \geq M^2,\ x^0 \gtrless 0 \right\}$, the wave equation being replaced by the Klein-Gordon equation

$$KG\,\Gamma \equiv (\square + M^2)\Gamma = 0 .$$

The variable $\varkappa$ can be replaced by a set $\vec{\varkappa}$ of space-like variables and the evenness requirement by that of invariance under the orthogonal group in $\vec{\varkappa}$ space.

Sketch of proof

(i) From C to Γ

Let (x, τ) be the variables conjugate to (p, κ). The product $C(x) \cos \kappa\sqrt{x^2}$ exists as a distribution in (x, κ), even in κ, which is infinitely differentiable in κ, polynomially bounded in κ, and coincides with C for $\kappa = 0$, because $\cos \kappa\sqrt{x^2}$ is polynomially bounded together with all its derivatives with respect to x, on the support of C, and the support of C is regular. Γ is defined as the Fourier transform with respect to x of $C(x) \cos \kappa\sqrt{x^2}$ and obviously is an even solution of the wave equation.

(ii) From Γ to C

Since Γ fulfils the wave equation, its Fourier transform $\tilde{\Gamma}$ has its support within the union of two opposite light cones in (p, κ) space. Hence, from the regularity property of this support, one can decompose $\tilde{\Gamma}$ according to

$$\tilde{\Gamma} = \tilde{\Gamma}^{+} - \tilde{\Gamma}^{-}$$

$\tilde{\Gamma}^{\pm}$ having its support on the $\begin{cases}\text{positive}\\ \text{negative}\end{cases}$ closed light cone. Γ can thus be written as the difference of boundary values of two polynomially bounded analytic functions holomorphic in tubes with these two self-dual cones as imaginary bases. The restrictions of these analytic functions to the analytic hyperplane $\kappa = 0$ exist, are polynomially bounded and analytic, in the remaining variables, in tubes with light cone basis, hence the support properties of the Fourier transforms of their boundary values.

B) - We now wish to take into account support properties of $\tilde{C}$, in p space, and derive from it support properties of Γ.

B.1. - The double cone property

If $\tilde{C}$ vanishes in a real neighbourhood of a time-like segment :

$$p = p_0 + t n \qquad n^2 > 0 \qquad -1 \leq t \leq +1$$

Γ vanishes in the double cone subtended by it :

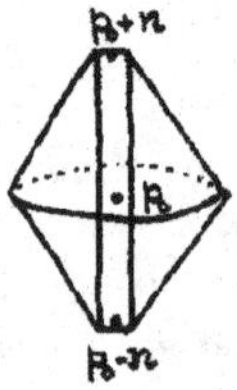

Using the wave equation and the evenness of Γ, one finds that Γ is zero together with all its derivatives where $\tilde{C}$ vanishes. One can then use the decomposition of Γ as a difference of boundary values of functions holomorphic in opposite tubes and the Kolm-Nagel version of the edge-of-the-wedge theorem [1], together with a completion by disks belonging to hyperbolae going through the end points of the time-like segment and asymptotic to light-like directions as indicated on Fig. 1.

One thus gains points by application of the continuity theorem, up to completely filling in the expected double cone. This is a geometric version of a proof first constructed by Borchers [5].

B.2. - Huyghens principle

Knowing some support properties for Γ, one solves the Cauchy problem on a hypersurface best adapted to the support of Γ : let Σ be a hypersurface which is spacelike, so that it is compactly intersected by an arbitrary light cone (one may even allow lightlike pieces provided they are bounded). Let Σ_+, Σ_- be the closed half-spaces defined by Σ. One can decompose Γ according to

$$\Gamma = \Gamma_+ - \Gamma_-$$

$\Gamma_\pm$ having support in $\Sigma_\pm$. Then

$$\Box\Gamma = \Box\Gamma_+ - \Box\Gamma_- = 0$$

Hence

$$\Box\Gamma_+ = \Box\Gamma_- = \sigma$$

where σ has its support on Σ. If a different decomposition is chosen, σ changes into $\sigma + \Box\tau$, where τ has its support on Σ. We call σ mod $\Box\tau$ the Cauchy data of Γ on Σ. They are distributions with support Σ defined modulo the Dalembertian of distributions with support Σ, and one can check that, locally where Σ is a manifold, this notion coincides with the usual one. If Γ vanishes in some open set Ω, it is clear that one can choose σ to vanish in $\Omega \cap \Sigma$. Let now D_{ret}, D_{adv} be the elementary retarded and advanced solutions of the wave operator. One can solve the inhomogeneous wave equations :

$$\Gamma_{\pm} = \mathring{\Gamma}_{\pm} + D_{\substack{ret\\adv}} * \sigma$$

where $\mathring{\Gamma}_{\pm}$ are solutions of the homogeneous wave equation which, under the geometrical assumptions we made, have supports $\Sigma_{\pm}$, bounded from below or above, hence identically zero. We thus obtain the representation

$$\Gamma = \Gamma_{+} - \Gamma_{-} = D * \sigma$$

where $D = D_{ret} - D_{adv}$ vanishes outside the union of two opposite light cones. As expected, Γ only depends on its Cauchy data σ mod $\Box\tau$, since $\Box D = 0$. The representation of Γ is able to yield further support properties : Γ vanishes outside the future and past shadows of the support of its Cauchy data (Huyghens principle), an information which has to be combined with the double cone property.

A more refined property [6] is the following. D is real analytic except on the skin of the light cone. Hence, if Γ vanishes on an open subset ω of a connected region $\mathcal{R}$ where no characteristic from a point of the support of σ goes, it vanishes identically in $\mathcal{R}$ (cf. Fig. 2).

Whereas the double cone and Huyghens principle are enough to obtain complete information when the input is the vanishing of $\tilde{C}$ between two space-like surfaces (the problem solved by JLD [3]), the analyticity principle was applied by Greenberg [6] to cases where the region of vanishing of $\tilde{C}$ is non connected.

As is well known, the JLD problem is connected with the following edge-of-the-wedge problem : let us decompose C into

$$C = R - A$$

Support $R = \bar{V}_{+}$, support $A = \bar{V}_{-}$.

Then $\tilde{R}$, $\tilde{A}$ are boundary values of functions analytic in opposite tubes, and the statement

$$\tilde{C} = 0 \qquad \text{in } \Omega$$

is converted into

$$\text{b.v. } \tilde{R} = \text{b.v. } \tilde{A} \qquad \text{in } \Omega ,$$

a statement to be exploited by use of the edge-of-the-wedge theorem and analytic completion. It turns out that all analyticity properties derivable by the JLD procedure could be obtained by analytical geometric methods including those which require the use of the analyticity principle. Some known results can only be obtained by geometrical methods [7]. Whatever analyticity follows from the JLD theorem can be obtained as follows : D can be decomposed in a non-unique way as the difference of boundary values of two functions analytic in tubes :

$$D = D^{(+)} - D^{(-)}$$

It turns out, however, that these functions have as sole singularities the points belonging to the hyperboloids

$$p^2 - k^2 = \rho \quad , \qquad \rho \geq 0 \; , \; \rho \text{ real} .$$

A careful choice of the decomposition of D allows to express the corresponding analytic functions Γ^+, Γ^- as

$$\Gamma^{\pm} = D^{(\pm)} * \sigma$$

which yields the analyticity described in Section 2, after investigation of the support of σ. This way, the JLD procedure solves an analytic problem "with growth", whereas the geometrical methods solve it without growth [7].

2. - PROBLEMS OF ANALYTIC COMPLETION [7)]

The analytic result that stems from JLD's analysis can be summarized as follows. Let $\mathcal{T}^{\pm}$ be the tubes with imaginary basis $V^{\pm}$, $\mathcal{R}$ a coincidence region, open, connected, bounded by two spacelike surfaces. The holomorphy envelope of the associated edge-of-the-wedge problem ($f^{\pm}$ holomorphic in $\mathcal{T}^{\pm}$, $\mathrm{bvf}^{+} = \mathrm{bvf}^{-}$ in $\mathcal{R}$) is the complement of the set of complex points of real "admissible" hyperboloids $\mathcal{H}_{u,\kappa}$:

$$(p-u)^2 = \kappa^2 \qquad u, \kappa \text{ real}, \ \kappa^2 \geq 0$$

which do not intersect $\mathcal{R}$.

We now look at a few special cases.

A) - $\mathcal{R}$ is unbounded from above (cf. Fig. 3). The only admissible hyperboloids are spacelike planes which do not intersect $\mathcal{R}$.

Hence $\mathcal{R}$ has to be convex ("re-entrant nose" phenomenon). The domain of analyticity is made up with :

- the real points of the convex hull of $\mathcal{R}$;
- all complex points of the real straight lines which intersect $\mathcal{R}$ (no such point lies in a real plane not intersecting $\mathcal{R}$.)

B) - The previous result can be applied to the following situation : $\mathcal{R}$ as above, but $V^{\pm}$ replaced by arbitrary (convex) cones.

Using two-dimensional hyperplanes and putting together all analyticity points deduced from A), one has the same result as in A). In particular, $V^{\pm}$ are enlarged to the $\begin{cases}\text{future}\\ \text{past}\end{cases}$ asymptotic cones of $\mathcal{R}$.

C) - Same as B) but $\mathcal{R} = \mathcal{R}^{+} \cup \mathcal{R}^{-}$ (cf. Fig. 4). The union of the domains pertaining to $\mathcal{R}^{+}$, $\mathcal{R}^{-}$ is natural and is the complement of the real and complex points of real hyperplanes intersecting neither $\mathcal{R}^{+}$ nor $\mathcal{R}^{-}$ (Hahn-Banach).

Remark

All these domains are of the following type : given an open set Ω in real space, the set of complex points of straight lines intersecting Ω, together with the real points of Ω is natural if Ω is convex. If the real points of Ω are deleted, one obtains a generalization of two opposite tubes.

From this point of view, the re-entrant nose phenomenon is not essentially different from the double cone phenomenon : in two dimensions let $\mathcal{R} = \Gamma_a^+ \cup \Gamma_b^+$ (cf, Fig. 5)

Within the domain pertaining to Γ_a^+, take the complex points of straight lines intersecting αa : make a real projective transformation which takes $a\alpha$ into the straight line at ∞ and apply the double cone theorem to a half-straight line in Γ_b^+. One could, of course, also apply the continuity theorem to a family of straight lines parallel to ab, starting high enough.

D) - A more general application [7] allows to describe in geometrical terms the holomorphy envelope associated with two arbitrary double cones in two dimensions, combining projective transformations and inversions [see E)] which transform finite double cones into infinite double cones. For instance, in the example shown on Fig. 6, the JLD result can be applied to situation II. There, the union of the two holomorphy envelopes is natural. In situation I, the adapted admissible hyperbolae go through the apex of the infinite cone. The solution to this problem was found during an attempt to understand the analytic equivalent to the analyticity principle in the JLD approach [7].

E) - $\Gamma^{+,(-)}$ is a topological product of $V^{+(-)}$'s, $\mathcal{R} = \Gamma_a^+$. The inversion

$$p \to p' = \{p'_i\} \qquad p'_i = b_i - \frac{p_i - b_i}{(p_i - b_i)^2}$$

transforms Γ_a^+ into the double cone of diagonal ab if b is in the past of a. It preserves the tubes of imaginary basis $\Gamma^{+,(-)}$. It transforms straight lines

$$p_i = u_i + v_i t \ , \quad u_i, v_i \text{ real} \ , \quad t \in \mathbb{C}$$

into curves which were called Q curves. In the JLD case, the Q curves are hyperbolae which allow to describe the domain from inside. The JLD domain consists of all complex points of doubly inadmissible hyperbolae, i.e., real hyperbolae asymptotic to the light cone with an upper sheet and a lower sheet which both intersect $\mathcal{R}$, together with its suitable completions (by double cone and re-entrant nose enlargements).

In the more general case of a double cone, one describes the domain in terms of complex points of doubly inadmissible Q curves. The description of the domain from outside, in more general cases, is due to Epstein and Glaser.

3. - SOME PROBLEMS OF GLOBAL DECOMPOSITIONS

All the problems which are dealt with here are solved with the help of the JLD theorem : they are problems "with bounds" but no solution is so far known without bound. They are all connected with the structure of the p space n point function of quantum field theory [2] summarized in these proceedings.

The starting point is the decomposition of the generalized retarded functions associated with a given channel ("hypercell") in terms of Bros trees [2]. The next step is to use the discontinuity formula for neighbouring geometric cells : the notations being the same as in the lecture by Epstein, Glaser and Stora (this volume), the discontinuity formula

$$\Delta_{\mathcal{S}_I} = t_{\mathcal{S}_I^+} - t_{\mathcal{S}_I^-} = \left(\Omega, \left[T_{\Sigma_{I'}}(I'), T_{\Sigma_I}(I)\right]\Omega\right)$$

contains both x space and p space support information, thus suggesting the use of the JLD theorem. Similar problems were first treated by Streater [8] in the context of the structure of Wightman functions. From Bros' tree decomposition, one sees that the difference $t_{\mathcal{S}_I^+} - t_{\mathcal{S}_I^-}$ involves only trees for which one of the defining equations is $S_I > 0$ or $S_I < 0$. Thus, one has the following situation :

$$\tilde{\Delta}_{\mathcal{S}_I} = \underset{p_I = 0}{b.v.}\left(\sum_{i \in J^+} \tilde{f}_i^{\,+} - \sum_{i \in J^-} \tilde{f}_i^{\,-}\right) = 0 \,, \quad p_I^2 < M_I^2$$

where the $f_i^{\pm}$ are holomorphic in simplicial tubes with imaginary basis contained in $\mathrm{Im}\,p_I \in V^{\pm}$ and M_I is a mass associated with the spectrum properties of the theory. Analyticity is furthermore deduced from x space support properties.

Let us choose a set of variables, one of which we denote ξ, $\xi = x_i - x_j$ where $i \in I$, $j \in I'$. By JLD, Δ_{ξ_I} is the restriction to $\kappa = 0$ of a distribution Γ in all variables together with κ, solution of the Klein-Gordon equation

$$KG\,\Gamma \equiv \left(\Box_\xi - \frac{\partial^2}{\partial\kappa^2} + M_I^2\right)\Gamma = 0$$

Using the double cone principle, we see that at least in the applications given here,

$$supp.\,\Gamma = supp.\,\Delta_{\xi_I} \times (-\infty < \kappa < +\infty)$$

because all other variables being fixed, the support in ξ is a union of $V^{\pm}$'s with different origins, whose complement contains arbitrary large timelike segments between the upper and lower boundary and all apices of V^+ cones are in the past of all apices of V^- cones. We can therefore split

$$\Gamma = \sum_{i \in J^+} \Gamma_i^+ - \sum_{i \in J^-} \Gamma_i^-$$

where

$$supp\,\Gamma_i^{\pm} = supp\,f_i^{\pm} \times (-\infty < \kappa < +\infty)$$

Hence

$$KG\,\Gamma = \sum_{i \in J^+} KG\Gamma_i^+ - \sum_{i \in J^-} KG\,\Gamma_i^- = 0$$

There exist therefore distributions $\Delta_{ij} = -\Delta_{ji}$, $i,j \in J^+ \cup J^-$ such that

$$\pm\, KG\,\Gamma_i^{\pm} = \sum_{j \in J^+ \cup J^-} \Delta_{ij}^{\pm\pm}$$

where

$$supp.\,\Delta_{ij}^{\pm\pm} = supp.\,\Gamma_i^{\pm} \cap supp\,\Gamma_j^{\pm}$$

hence

$$\Gamma_i^{\pm} = \mathring{\Gamma}_i^{\pm} + D_{\substack{ret\\adv}} * \sum_{j \in J^+ \cup J^-} \Delta_{ij}^{\pm\pm}$$

where the convolution exists, has its support contained in the support of $\Gamma_i^{\pm}$, hence, also that of $\mathring{\Gamma}_i^{\pm}$. This support being bounded from above or from below, $\mathring{\Gamma}_i^{\pm}$ vanishes identically. Putting pieces together, we get

$$\Gamma = D * \sum_{i \in J^+, j \in J^-} \Delta_{ij}^{+-}$$

with restriction to $\mathcal{K} = 0$ equal to $\Delta_{\mathcal{S}_1}$. A suitable decomposition $D = D^{ret} - D^{adv}$ yields a common analytic decomposition in p space, for

$$t_{\mathcal{S}_I^{\pm}} = \sum_{\substack{i \in J^+ \\ j \in J^-}} D^{\substack{ret\\adv}} * \Delta_{ij}^{+-}$$

It is of the form (in p space)

$$\tilde{t}_{\mathcal{S}_I^{\pm}} = \sum_{i \in J^+, j \in J^-} \tilde{f}_{ij}$$

where $\tilde{f}_{ij}$ is holomorphic in the tube convex hull of the tubes defining the trees f_i, f_j indented by the cut $p_I^2 = M_I^2 + \rho$, ρ real positive. The ambiguity of this decomposition is of the form

$$\delta \tilde{f}_{ij} = \sum_{k \in J^+ \cup J^-} \tilde{f}_{ijk}$$

where $\tilde{f}_{ijk}$ is the common analytic continuation of

$$\widetilde{D^{ret} * \Delta_{ijk}^{+-}} \quad , \quad \widetilde{D^{adv} * \Delta_{ijk}^{+-}} \, ,$$

the support of Δ_{ijk} being the intersection of those corresponding to trees f_i, f_j, f_k. Hence $\tilde{f}_{ijk}$ is analytic in the convex hull of the tubes defining f_i, f_j, f_k, indented by the cut $p_I^2 = M_I^2 + \rho$.

We now list a few applications of this decomposition property to low values of n. We shall not write down supports and their duals in detail but rather show the s space corresponding cell. We shall write equations involving cells, each cell symbol representing a function analytic in the corresponding Araki tube - wiggly lines represent cuts. In the following, E.W. means edge-of-the-wedge, JLD means Jost-Lehmann-Dyson.

A) - Three-point function. s space is shown in Fig. 7. Each couple of neighbouring cells gives rise to a holomorphy envelope [8] ;

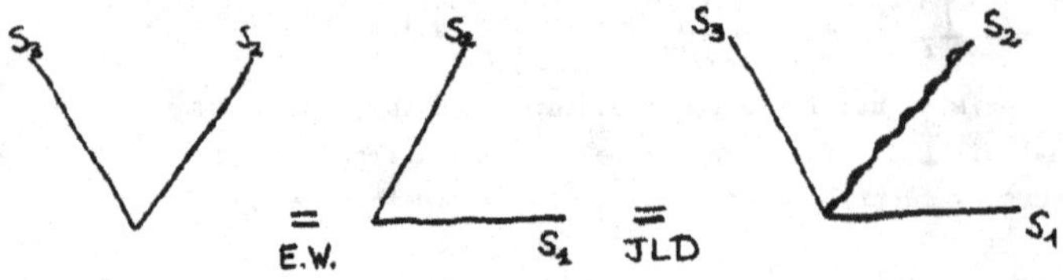

B) - Four-point function. s space is shown in Fig. 8.

1)

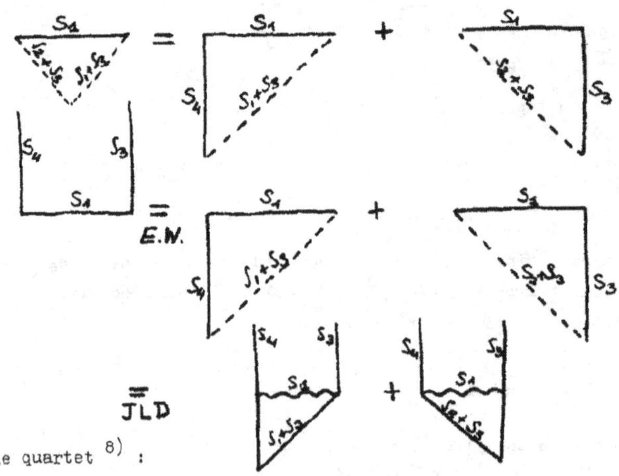

2) the quartet [8] :

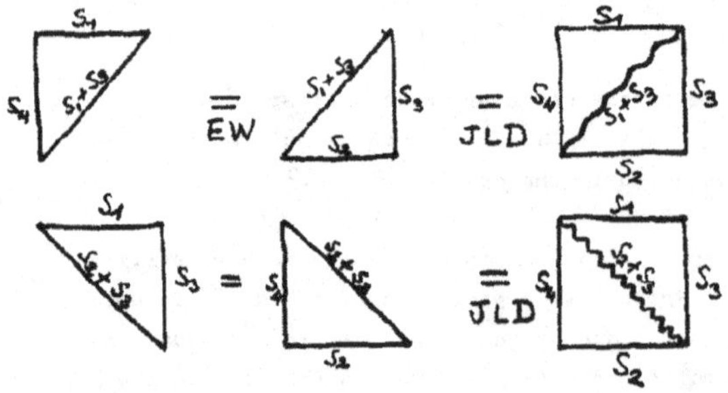

Hence the common decomposition for all four members of the quartet hypercell :

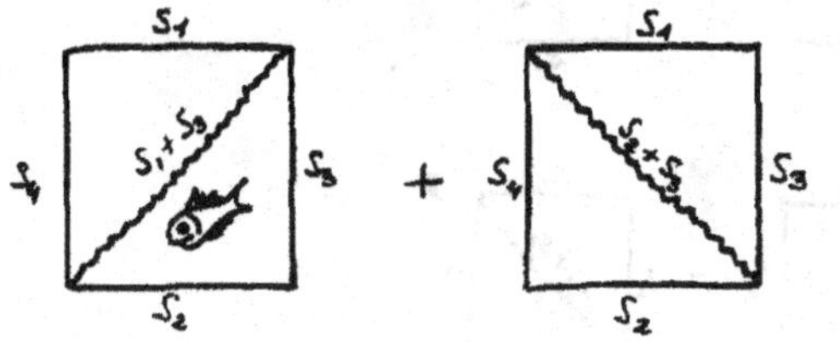

3) putting together both decompositions we get :

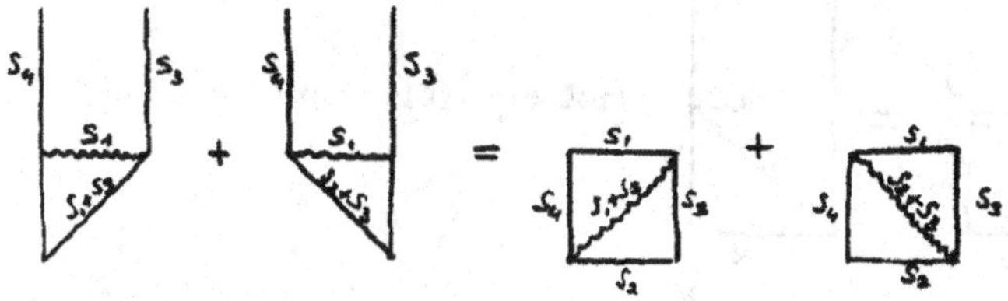

hence :

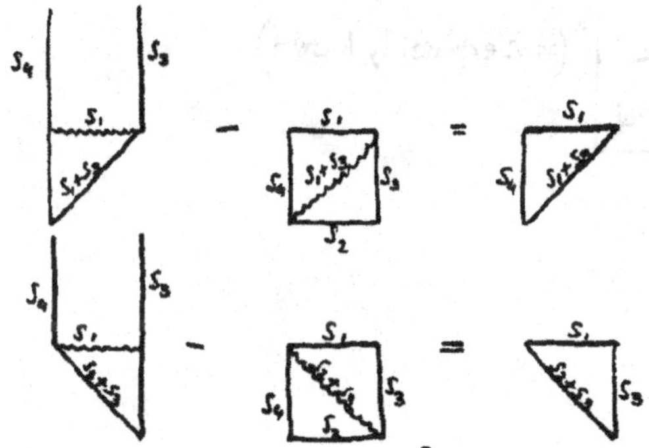

= holomorphy envelope

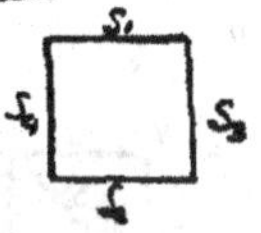

defining

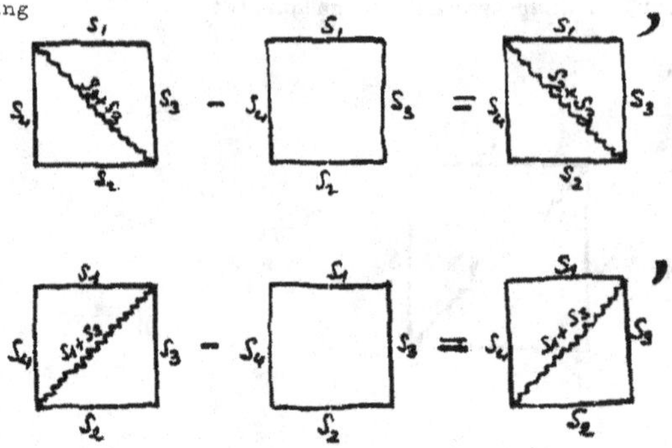

we get :

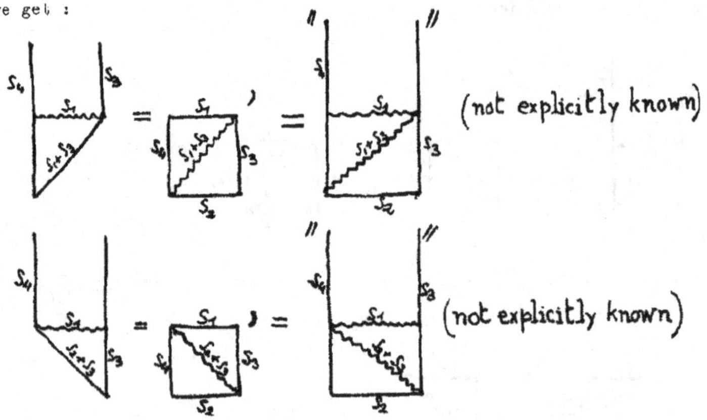

hence the global decomposition

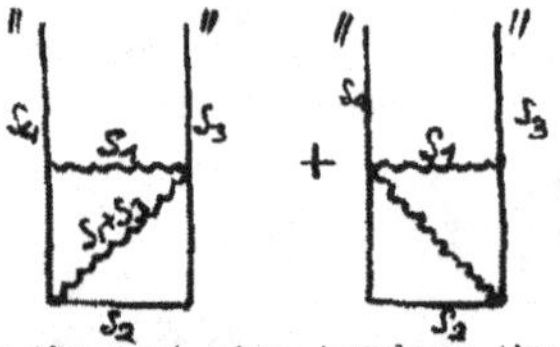

Unfortunately the next step involves the relationship

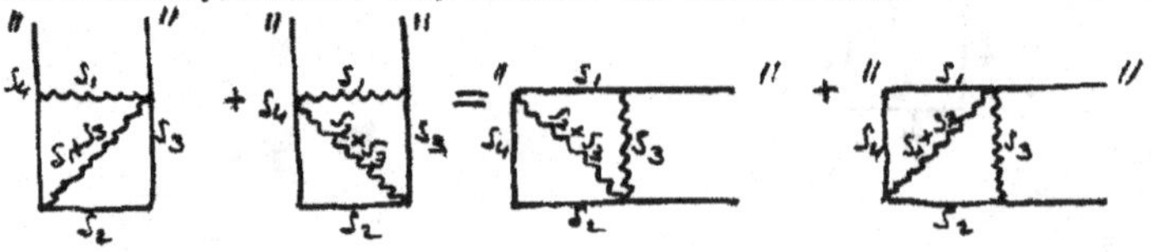

C) - Five-point function. Two types of Bros trees are shown in Fig. 9 : 
ABCD, BCDE ; there are six of each kind within the hypercell represented by ABCDEF (cf. Fig. 7). Their convex hull ABCDE indented by the cut BCD will be denoted ABCDE. On our drawing, there is another similar object : FABDC. Let (EF) denote the holomorphy envelope of these two domains (which is not explicitly known). There are six such domains, each one corresponding to a diagonal of a vertical facet. In terms of these six functions, one has the global decomposition, valid for the analytic continuation of any member of the hypercell :

$$F = (AD) + (CF) + (AE) + (BC) + (EF) + (BD)$$

CONCLUSION AND OUTLOOK

The partial answers exhibited here pose the more general following problems :

- For general n, can one reduce the problem which arises from the Steinmann identities within a hypercell to the construction of well-defined holomorphy envelopes ?
- When looking at neighbouring hypercels, one is faced with problems of the following type : let f_i be holomorphic in Ω_i, $i \in I$, let

$$\sum_i f_i = 0 \quad \text{in} \quad \bigcap_{i \in I} \Omega_i$$

what are the implications of such identities ? Do these problems belong to some part of cohomology theory ?
- In other words, can the problem posed by the structure of the n point p space function of quantum field theory be reduced to the construction of well-defined holomorphy envelopes ?

REFERENCES

This list is not exhaustive, but mostly gives a sample of references where more bibliography can be found, besides few originals which are not very well known.

1) "Hyperfunctions and Pseudo-Differential Equations", Lecture Notes in Mathematics, Vol. 287, Springer Verlag (1973) ;
"Théorie des Hyperfonctions", by P. Shapira, Lecture Notes in Mathematics, Vol. 126, Springer Verlag (1970) ;
A. Martineau - Lectures at the Meetings of RCP 25, Strasbourg Mathematics Institute, Vol. 3 (1967).

2) J. Bros and R. Stora - Lectures at the Meetings of RCP 25, Strasbourg Mathematics Institute, Vol. 3 (1967) ;
H. Epstein - Brandeis Summer Institute (1965) ; Gordon and Breach, New York (1966) ;
H. Epstein, V. Glaser and R. Stora - These Proceedings.

3) A.S. Wightman - Lectures at Les Houches Summer School of Theoretical Physics (1960) ; Hermann, Paris (1961) ;
V.S. Vladimirov - "Methods of the Theory of Functions of Several Complex Variables", Cambridge M.I.T. Press (1966).

4) R. Stora - unpublished (1964), summarized in Lectures at the Meetings of RCP 25, Strasbourg Mathematics Institute, Vol. 2 (1966).

5) H. Borchers - Nuovo Cimento 19, 787 (1961).

6) O.W. Greenberg - J.Math.Phys. 3, 859 (1962).

7) J. Bros, A. Messiah and R. Stora - J.Math.Phys. 2, 639 (1961) ;
H. Borchers and R. Stora - unpublished (1962).
For another type of solutions with growth, see :
R. Seneor - Commun.Math.Phys. 11, 233 (1968).

8) R.F. Streater - Proc.Roy.Soc. A256, 39 (1960).

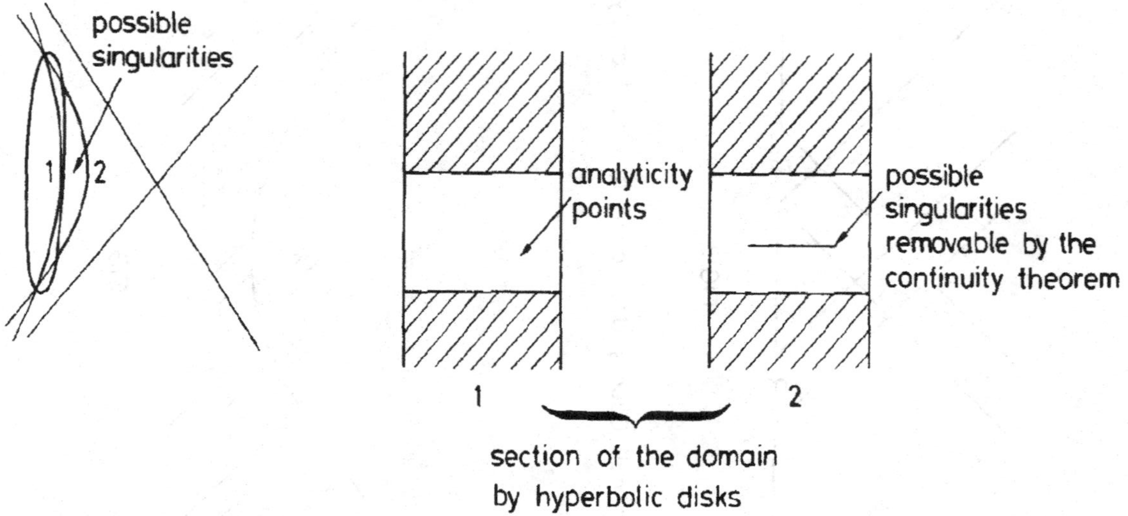

FIG.1 Analytic proof of the double cone theorem.

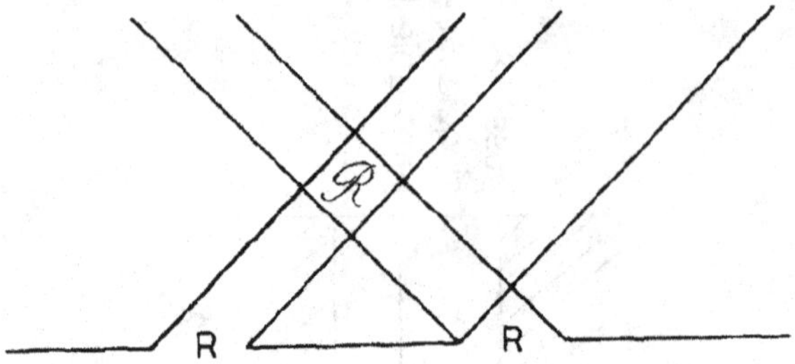

FIG. 2

If I has null Cauchy data on R it is real analytic in $\mathscr{R}$.

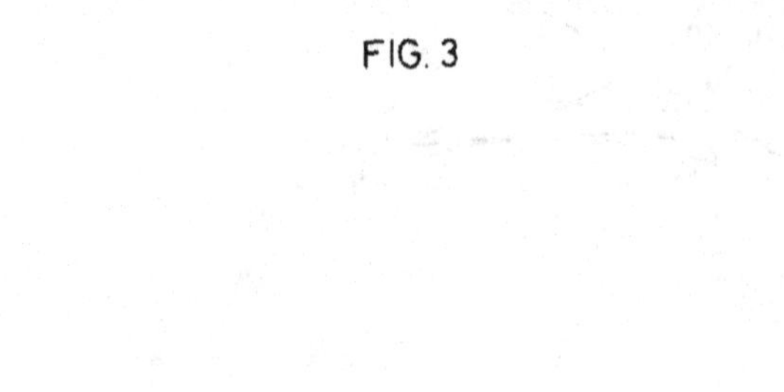

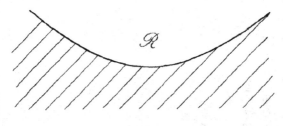

FIG. 3

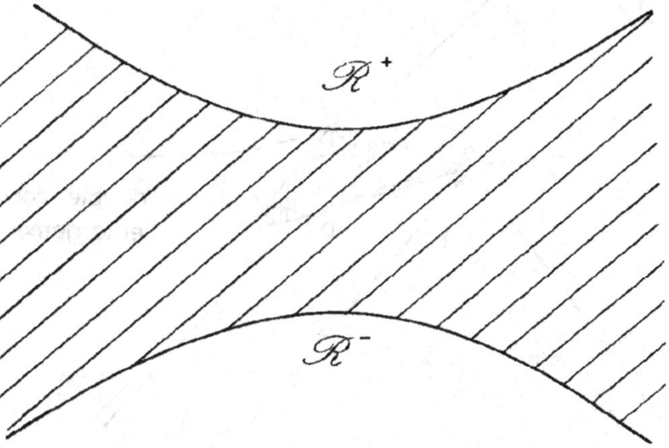

FIG. 4

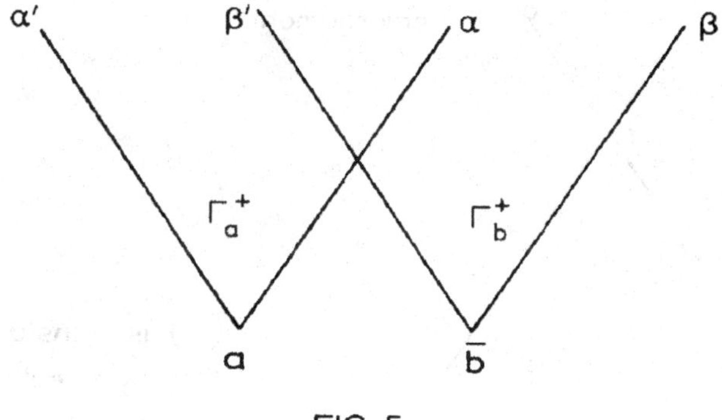

FIG. 5

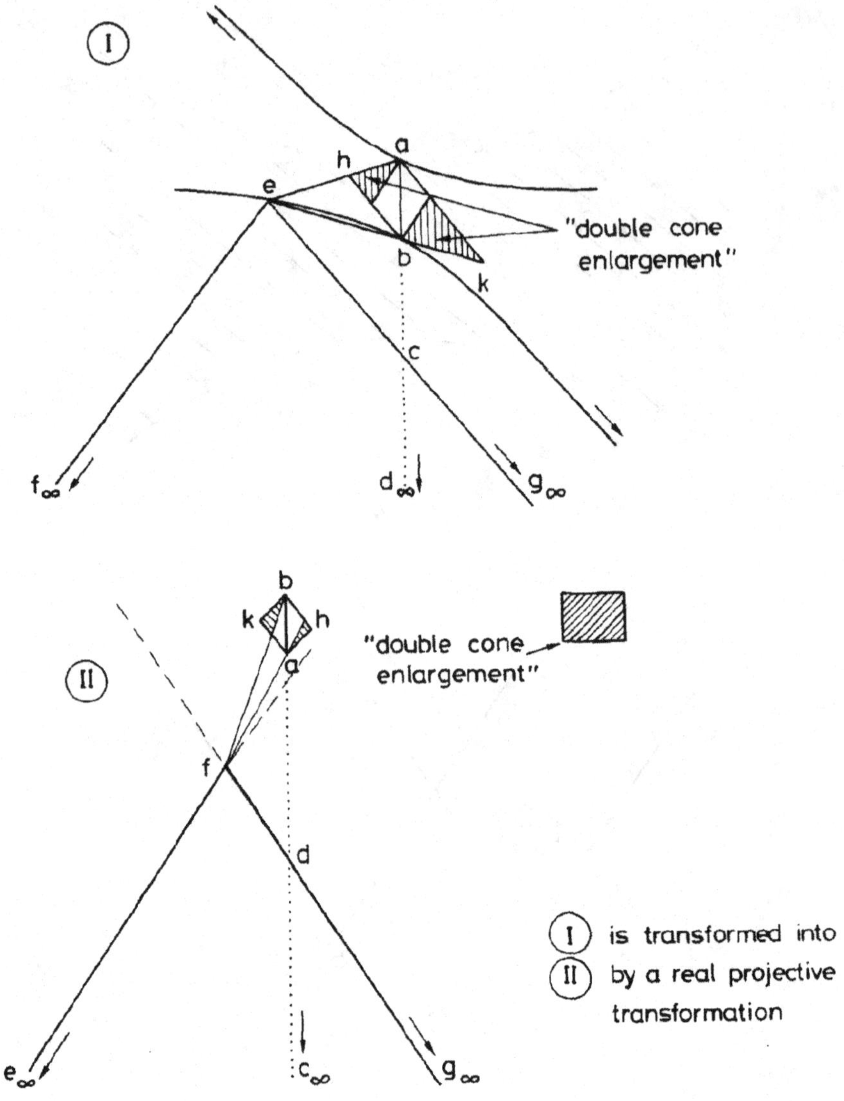
I
a
h
e
"double cone
enlargement"
b
k
c
f∞
d∞
g∞
b
k
h
"double cone
enlargement"
a
II
f
d
I is transformed into
II by a real projective
transformation
e∞
c∞
g∞

FIG.6

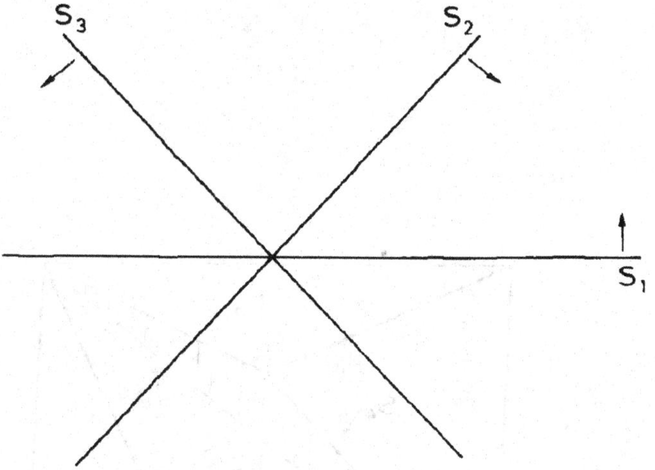

FIG.7

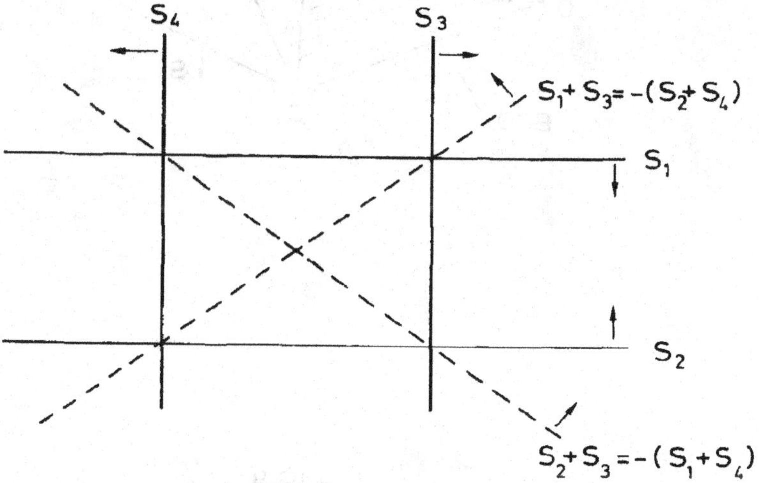

FIG. 8

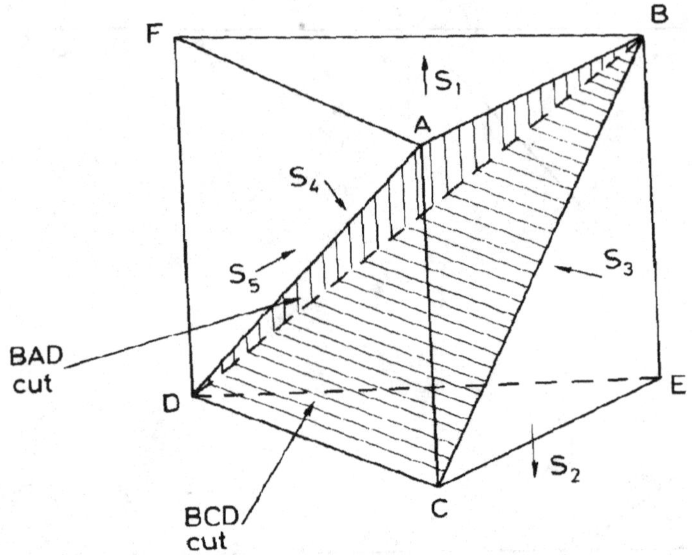

FIG. 9

QUELQUES ASPECTS GLOBAUX DES PROBLEMES D'EDGE - OF - THE - WEDGE

J. BROS, H. EPSTEIN, V. GLASER et R. STORA

Contribution au Colloque sur les Hyperfonctions et leurs Applications,
Nice, Mai 1973 - (Présentée par H. EPSTEIN)

Cet exposé décrit brièvement quelques résultats obtenus entre 1961 et 1963. Certains d'entre eux ont été publiés dans [1], [2], [3]. D'autres n'ont pas encore été publiés. L'accent est mis sur l'aspect global des problèmes du type d'edge-of-the-wedge. L'exploitation "locale", ou, plus exactement, infinitésimale, du théorème e.o.w. en théorie des champs est en effet presque achevée [5] cependant l'utilisation de la "positivité" pourrait ouvrir de nouvelles possibilités; (voir l'exposé de Glaser) ; tout au contraire les résultats globaux connus sont très rares et hétéroclites. Cela est d'autant plus regrettable que le plus célèbre d'entre eux (dont il ne sera pas question ici ; voir l'exposé de Stora), dû à Jost, Lehmann et Dyson (JLD) [6,7] s'est montré extrêmement utile dans un grand nombre de problèmes de la théorie des champs. On commencera par traiter un exemple très simple dont on peut déduire une étonnante variété d'applications. Sauf exceptions, toutes les valeurs aux bords seront prises au sens des fonctions C^{∞}, l'extension aux distributions et hyperfonctions ne posant, en général, pas de problème majeur.

I.- EXEMPLE SIMPLE.

Variables : $z_1 = x_1 + iy_1 \in \mathbb{C}$, $z_2 = x_2 + iy_2 \in \mathbb{C}$.

1°) Soient deux fonctions :

$f_1(z_1,x_2)$, C^∞ dans $\{x_1,y_1,x_2 : 0 \le y_1 \le \pi\}$, analytique en

z_1 pour $0 < y_1 < \pi$;

$f_2(x_1,z_2)$, C^∞ dans $\{x_1,x_2,y_2 : 0 \le y_2 \le \pi\}$, analytique en

z_2 pour $0 < y_2 < \pi$.

2°) On suppose de plus que, dans leur domaine de définition, ces fonctions satisfont à :

$$|f_j(z_1,z_2)| \le C\,(1+|z_1|)^{-1}\,(1+|z_2|)^{-1} .$$

3°) Enfin on suppose que :

$f_1(x_1,x_2)=f_2(x_1,x_2)$ pour tout $(x_1,x_2) \in \mathbb{R}^2$ tel que

$|x_1+x_2| < \varepsilon$ (où ε est > 0) .

Dans ces conditions

<u>Lemme 1</u> : il existe F holomorphe dans H , et C^∞ dans $\overline{H}$, où

$$H = \{(z_1,z_2) \in \mathbb{C}^2 : 0 < y_1, 0 < y_2, y_1 + y_2 < \pi\} ,$$

telle que $F(z_1,x_2) = f_1(z_1,x_2)$ pour $0 \leq y_1 \leq \pi$, $x_2 \in \mathbb{R}$, $F(x_1,z_2) = f_2(x_1,z_2)$ pour $0 \leq y_2 \leq \pi$, $x_1 \in \mathbb{R}$; en particulier si $(x_1, x_2) \in \mathbb{R}^2$, on a $F(x_1,x_2) = f_1(x_1,x_2) = f_2(x_1,x_2)$.

<u>Démonstration</u>

Posons, pour $z = (z_1, z_2) \in H$,

$$F(z_1,z_2) = \frac{1}{2\pi i} \int_{-\infty}^{\infty} \frac{f_1(z_1+z_2 - \theta_2, \theta_2)}{\theta_2 - z_2} d\theta + \frac{1}{2\pi i} \int_{-\infty}^{\infty} \frac{f_2(\theta_1, z_1+z_2-\theta_1)}{\theta_1 - z_1} d\theta_1$$

Des raisonnements classiques montrent que $F \in \mathcal{O}(H)$ et que F est C^∞ sur $\overline{H}$. Choisissons $(x_1, x_2) \in \mathbb{R}^2$ tel que $|x_1+x_2| < \epsilon$. Alors

$$F(x_1,x_2) = \lim_{\substack{\eta\to 0\\ \eta>0}} \left[\frac{1}{2\pi i} \int \frac{f_1(x_1+x_2 - \theta_2 + 2i\eta , \theta_2)}{\theta_2-x_2-i\eta} d\theta_2 + \frac{1}{2\pi i} \int \frac{f_2(\theta_1, x_1 + x_2 - \theta_1 + 2i\eta)}{\theta_1-x_1-i\eta} d\theta_1\right]$$

$$= \lim_{\substack{\eta\to 0\\ \eta>0}} \left[\frac{1}{2\pi i} \int \frac{f_1(x_1 - \theta + 2i\eta , x_2 + \theta)}{\theta - i\eta} d\theta - \frac{1}{2\pi i} \int \frac{f_2(x_1-\theta, x_2 + \theta + 2i\eta)}{\theta + i\eta} d\theta\right]$$

et il est facile de voir que cette limite est égale à

$$\lim_{\substack{\eta\to 0\\ \eta>0}} \frac{1}{2\pi i}\int\left[\frac{f_1(x_1-\theta\,,\,x_2+\theta)}{\theta-i\eta} - \frac{f_2(x_1-\theta\,,\,x_2+\theta)}{\theta+i\eta}\right]d\theta \quad .$$

Puisque $|(x_1-\theta)+(x_2+\theta)| = |x_1+x_2| < \varepsilon$, ceci est égal à $f_1(x_1,x_2) = f_2(x_1,x_2)$. Par prolongement analytique il en découle $f_1(z_1,x_2) = F(z_1,x_2)$ pour $0 \le y_1 \le \pi$, $x_2 \in \mathbb{R}$, et $f_2(x_1,z_2) = F(x_1,z_2)$ pour $x_1 \in \mathbb{R}$, $0 \le y_2 < \pi$. Le lemme est démontré.

Introduisons maintenant une transformation conforme très utile : l'application

$\zeta \to z = \log \frac{\tau+\zeta}{\tau-\zeta}$ (où τ est >0) , (dont l'inverse est $\zeta = \tau \operatorname{th} \frac{z}{2}$) , applique biunivoquement le plan coupé

$$\{\zeta : \operatorname{Im}\zeta \neq 0 \text{ ou } |\zeta| < \tau\} =$$

$$= \{\zeta : -\pi < \operatorname{Arg}\frac{\tau+\zeta}{\tau-\zeta} < \pi\} ,$$

sur la bande

$$\{z : -\pi < \operatorname{Im} z < \pi\} .$$

En particulier l'image du domaine (lunule circulaire)

$$D(\tau,\alpha) = \{\zeta \in \mathbb{C} : 0 < \operatorname{Arg}\frac{\tau+\zeta}{\tau-\zeta} < \alpha\}$$

(où $0 < \tau$, $0 < \alpha \le \pi$) est la bande

$$\{z \in \mathbb{C} : 0 < \operatorname{Im} z < \alpha\}$$

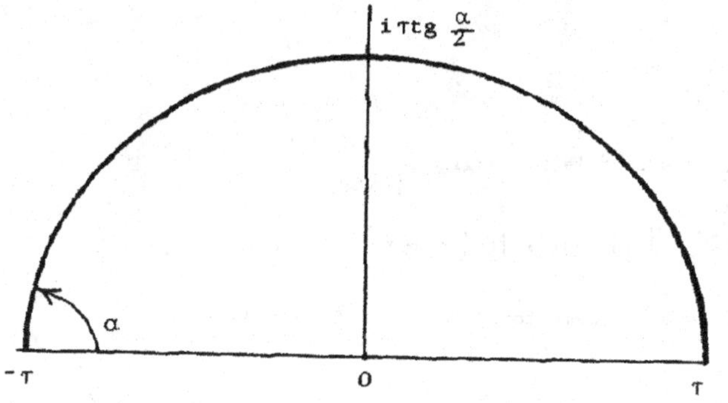

Fig.1 : le domaine $D(\tau,\alpha)$.

ette remarque nous permet de "localiser" le lemme 1 :

otons $\zeta_1 = \xi_1 + i\eta_1$, $\zeta_2 = \xi_2 + i\eta_2$ deux variables complexes.

oient φ_1 et φ_2 deux fonctions de ζ_1 et ζ_2 telles que :

°) $\varphi_1\ (\zeta_1 , \zeta_2)$ est définie et C^∞ quand $\zeta_2 = \xi_2 \in \mathbb{R}$ et
$_1 \in \overline{D}(\tau_1,\alpha_1)$, et analytique en ζ_1 dans $D(\tau_1, \alpha_1)$ pour chaque
$_2 = \xi_2$ réel ; (on suppose $0 < \tau_1$, $0 < \alpha_1 \leq \pi$) ;

°) φ_2 a les mêmes propriétés en échangeant ζ_1 et ζ_2 et en
emplaçant τ_1 , α_1 par τ_2 , α_2 ;

°) Il existe dans $\mathbb{R}^2$ un voisinage V de

$$\{(\xi_1,\xi_2) \in \mathbb{R}^2 ; \frac{\xi_1}{\tau_1} = -\frac{\xi_2}{\tau_2} \ , \ -\tau_1 < \xi_1 < \tau_1\}$$

el que $\varphi_1\ (\xi_1 , \xi_2) = \varphi_2\ (\xi_1 , \xi_2)$ dans V.

Dans ces conditions :

Lemme 2

φ_1 et φ_2 coïncident dans tout le rectangle

$$\{\xi = (\xi_1 , \xi_2) \in \mathbb{R}^2 : |\xi_1| < \tau_1 , |\xi_2| < \tau_2\}$$

et sont les valeurs aux bords d'une même fonction $G(\zeta_1,\zeta_2)$ analytique dans

$$H_1 = \{\zeta = (\zeta_1,\zeta_2) : 0 < \text{Im}\,\zeta_1 , 0 < \text{Im}\,\zeta_2 ,$$

$$\frac{1}{\alpha_1} \text{Im}\, \frac{\tau_1 + \zeta_1}{\tau_1 - \zeta_1} + \frac{1}{\alpha_2} \text{Im}\, \frac{\tau_2 + \zeta_2}{\tau_2 - \zeta_2} < 1\}$$

et C^∞ dans $\overline{H}_1$.

Démonstration

En remplaçant τ_j par $\tau_j' = (1-\varepsilon)\,\tau_j$, $(j = 1,2 \; ; \; 0 < \varepsilon < 1)$ on est ramené au cas où V contient l'ouvert réel :

$$W = \{\xi_1, \xi_2 : |\xi_j| < \tau_j' \; (j=1,2), \; \left|\frac{\xi_1}{\tau_1'} + \frac{\xi_2}{\tau_2'}\right| < K\}$$

où $0 < K$. Considérons les transformations conformes:

$$z_j = \frac{\pi}{\alpha_j} \log \frac{\tau'_j + \zeta_j}{\tau'_j - \zeta_j} \qquad (j = 1,2) \quad ,$$

$$\zeta_j = \tau'_j \ \text{th} \ \frac{\alpha_j z_j}{2\pi} \quad .$$

Quand $z_1 = x_1$, $z_2 = x_2$, réels, on a

$$\left|\text{th} \frac{\alpha_1 x_1}{2\pi} + \text{th} \frac{\alpha_2 x_2}{2\pi}\right| \leq 2 \left|\text{th} \frac{\alpha_1 x_1 + \alpha_2 x_2}{2\pi}\right| < \frac{|\alpha_1 x_1 + \alpha_2 x_2|}{\pi} \quad .$$

Donc la bande

$$\{(x_1, x_2) \in \mathbb{R}^2 \ : \ |\alpha_1 x_1 + \alpha_2 x_2| < \pi K\} \quad ,$$

est dans l'image de W . De plus les fonctions

$$f_1(z_1, x_2) = (z_1 + i)^{-1} (x_2 + i)^{-1} \varphi_1(\tau'_1 \ \text{th} \frac{\alpha_1 z_1}{2}, \ \tau'_2 \ \text{th} \frac{\alpha_2 x_2}{2})$$

$$f_2(x_1, z_2) = (x_1 + i)^{-1} (z_2 + i)^{-1} \varphi_2(\tau'_1 \ \text{th} \frac{\alpha_1 x_1}{2}, \ \tau'_2 \ \text{th} \frac{\alpha_2 x_2}{2}) \quad ,$$

vérifient les hypothèses 1°) et 2°) du lemme 1. Il en est a fortiori de même pour

$$F_1(z_1, x_2) = f_1(\frac{\alpha_1}{\pi} z_1, \frac{\alpha_2}{\pi} x_2) \ , \ F_2 = f_2(\frac{\alpha_1}{\pi} x_1, \frac{\alpha_2}{\pi} z_2) \quad .$$

Le lemme 1 montre que ces dernières coïncident sur tous les réels ; donc f_1 et f_2 satisfont à toutes les conditions du lemme 1 . En revenant aux variables ζ_1 et ζ_2 , et en faisant tendre τ'_j vers τ_j , on en déduit le lemme 2. Remarquons qu'en faisant tendre τ_1

et τ_2 vers l'infini en maintenant $\tau_j \operatorname{tg} \frac{\alpha_j}{2}$ borné, on généralise aussi le lemme 1 au cas où $f_{1,2}$ ne sont pas nécessairement bornées à l'infini, et où, initialement, elles ne coïncident que sur un voisinage quelconque de la droite $\{x_1, x_2 : x_1 + x_2 = 0\}$. Les généralisations au cas où on a n fonctions $f_k(x_1, \ldots, x_{k-1}, z_k, x_{k+1}, \ldots, x_n)$, chacune analytique lorsque $\operatorname{Im} z_j = 0$ $(j \neq k)$, $0 < \operatorname{Im} z_k < \pi$, et coïncidant dans un voisinage de l'hyperplan $\{x = (x_1, \ldots, x_i) : \sum_{j=1}^{n} x_j = 0\}$ peut se faire sans difficulté : dans le cas borné, on part de la formule

$$F(z_1, \ldots, z_n) = \sum_{k=1}^{n} \frac{1}{(2i\pi)^{n-1}} \int f_k(\theta_1, \ldots, \theta_{k-1}, \theta_k + \sum_{j=1} z_j, \theta_{k+1}, \ldots, \theta_n) \times$$

$$\times \, \delta\Big(\sum_{j=1}^{n} \theta_j\Big) \prod_{r \neq k} (\theta_r - z_r)^{-1} \prod_{\ell=1}^{n} d\theta_\ell \; .$$

On continue en faisant les mêmes transformations conformes que pour $n = 2$.

Conclusions :

1) Le théorème du tube est applicable à des situations "aplaties" : c'est le résultat de MALGRANGE-ZERNER (1961).

2) Il y a des phénomènes intéressants d'agrandissement de la région de coïncidence dans les réels. Nous y reviendrons plus loin.

II.- LES PROBLEMES D'E.O.W. à 2 TUBES.

Ils se posent ainsi : deux fonctions f_1 et f_2 sont analytiques respectivement dans $\mathfrak{T}_1 = \mathbb{R}^n + i\, C_1$ et $\mathfrak{T}_2 = \mathbb{R}^n + i\, C_2$; C_1 et C_2 sont deux cônes convexes dans $\mathbb{R}^n$ qui peuvent être soit ouverts, soit "aplatis" ; les valeurs aux bords de f_1 et f_2 dans les réels coïncident dans un ouvert réel $\mathfrak{R}$ de $\mathbb{R}^n$ ("région de coïncidence"). Il s'agit de trouver (ou tout au moins d'étudier) le "domaine" $H(\mathfrak{T}_1 \cup \mathfrak{T}_2 \cup \mathfrak{R})$ où elles ont un prolongement commun. En pratique on rencontre essentiellement deux cas

a) $C_1 = - C_2$ (problèmes d'e.o.w. opposé)

b) $C_1 \cap (-C_2) = \emptyset$ (e.o.w. oblique) .

Il y a de plus des cas, importants en pratique, d'une nature "semi-locale": on a à trouver $H((\mathfrak{T}_1 \cup \mathfrak{T}_2 \cup \mathfrak{R}) \cap \Omega)$, où Ω est un voisinage ouvert complexe donné de $\mathfrak{R}$. Comme il a déjà été dit, il y a peu de résultats généraux. Mais il existe des exemples solubles assez intéressants.

II.1.- E.O.W. OPPOSE.

A. (Cônes polyédraux simpliciaux).

$$C_+ = \{y = (y_1,\dots,y_n) \in \mathbb{R}^n \ : \ y_1 > 0,\dots,y_n > 0\} = - C_-$$

$$\mathfrak{J}_\pm = \mathbb{R}^n \pm i\, C_+$$

$$\mathfrak{R} = \{x \in \mathbb{R}^n : |x_j| < 1 \quad \text{pour tout } j = 1,\dots,n\} \ .$$

Dans ces cas on peut facilement calculer $H(\mathfrak{J}_+ \cup \mathfrak{J}_- \cup \mathfrak{R})$: la transformation

$$\zeta_j = \log \frac{1 + z_j}{1 - z_j} \quad , \quad z_j = \text{th} \frac{\zeta_j}{2} \quad ,$$

remplace le problème par celui de trouver $H(\mathfrak{J}'_+ \cup \mathfrak{J}'_- \cup \mathbb{R}^n)$ où

$$\mathfrak{J}'_+ = - \mathfrak{J}'_- = \{\zeta \in \mathbb{C}^n : 0 < \text{Im } \zeta_j < \pi\} \ .$$

Vues les remarques faites à propos du lemme 1 et de ses généralisations, on peut appliquer le théorème du tube convexe, et on trouve

$$H(\mathfrak{J}'_+ \cup \mathfrak{J}'_- \cup \mathbb{R}^n) = \bigcup_{0 \le \theta \le \pi} \{\zeta : -\theta < \text{Im } \zeta_j < \pi - \theta\} \ .$$

Donc

$$H(\mathfrak{J}_+ \cup \mathfrak{J}_- \cup \mathfrak{R}) = \bigcup_{0 \le \theta \le \pi} \{z \in \mathbb{C}^n : -\theta < \text{Arg} \frac{1+z_j}{1-z_j} < \pi - \theta\} \ .$$

Le domaine peut donc être décrit comme une union de polydisques :

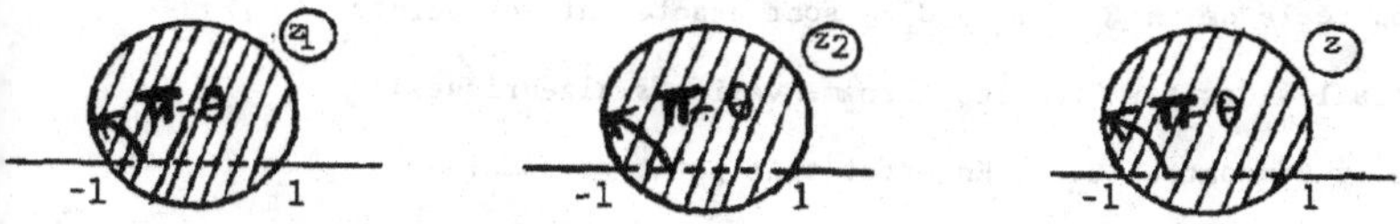

Fig.2 : $H(\mathcal{T}_+ \cup \mathcal{T}_- \cup \mathfrak{K})$ = union de polydisques.

Il est clair qu'on obtient de même, par un changement de variables réel-linéaire, le domaine $H(\mathcal{T}_+ \cup \mathcal{T}_- \cup C(a,b))$, où $C(a,b)$ est le double -cône (ici parallélépipède)

$$C(a,b) = \{x \in \mathbb{R}^n : x - a \in C_+ , b - x \in C_+\} .$$

B. $\mathfrak{K} = C_+$.

En particulier il est intéressant de considérer le cas où $a = 0$ et $b_j = +\infty$ pour tout j, c'est-à-dire le cas où $\mathfrak{K} = C_+$. Ce cas peut être ramené au théorème du tube en posant

$$\zeta_j = \log z_j \quad , \quad z_j = e^{\zeta_j}$$

et on trouve :

$$H(\mathcal{T}_+ \cup \mathcal{T}_- \cup C_+) = \bigcup_{0 \leq \theta \leq \pi} \{z \in \mathbb{C}^n : - \theta \leq \operatorname{Arg} z_j \leq \pi - \theta\}$$
$$= \bigcup_{0 \leq \theta \leq \pi} e^{-i\theta} \mathcal{T}_+ = \bigcup_{s : \operatorname{Im} s < 0} s\mathcal{T}^+ .$$

Il est utile de donner une autre représentation géométrique de ce domaine .

Les points non réels de $H(\mathfrak{J}_+ \cup \mathfrak{J}_- \cup C_+)$ sont exactement les points non réels de toutes les droites "réelles" (comme variétés algébriques) dont la trace réelle rencontre C_+ . En effet tout point du domaine s'écrit

$$z = (\alpha - i)\,(x_0 + iy_0)\ ,\ y_0 \in C_+,\quad \alpha \in \mathbb{R}$$

ou

$$z = (1 + \alpha^2)\, y_0 + (\alpha - i)\,(x_0 - \alpha\, y_0) \qquad ,$$

et l'équation

$$z = (1 + \alpha^2)\, y_0 + \tau\,(x_0 - \alpha\, y_0) \quad , \quad \tau \in \mathbb{C} \qquad ,$$

est bien celle d'une droite "à coefficients réels" dont la trace réelle rencontre C_+ (en particulier au point $(1 + \alpha^2)\, y_0$). La réciproque est aussi vraie.

C. $\underline{\mathfrak{K} = \hat{\mathfrak{K}} + C_+}$.

Considérons maintenant le cas où $\mathfrak{K}$ est une union de translatés de C_+ , i.e. où

$$\hat{\mathfrak{K}} + C_+ = \mathfrak{K} \qquad .$$

Danc ce cas

$$H(\mathfrak{J}_+ \cup \mathfrak{J}_- \cup \mathfrak{K}) \supset \bigcup_{a \in \mathbb{R}} \ \bigcup_{0 \leq \theta < \pi} (a + e^{-i\theta}\, \mathfrak{J}_+)$$

qui contient à son tour

$$\bigcup_{a \in \mathcal{R}} (a - i\mathcal{J}_+) = i\,\mathbb{R}^n + \mathcal{R} \quad .$$

L'enveloppe d'holomorphie de ce tube est son enveloppe convexe ; donc $\mathcal{R}$ n'est naturelle que si elle est convexe. En particulier, si on se donne initialement une région de coïncidence $\mathcal{R}$ non convexe (mais telle que $\mathcal{R} + C_+ = \mathcal{R}$) , elle s'agrandit automatiquement à son enveloppe convexe : c'est là un deuxième phénomène d'agrandissement de la région réelle de coïncidence. (Ce phénomène est lié à celui des "reentrent noses" découvert par Dyson [7]). Lorsque $\mathcal{R}$ est convexe et vérifie $\mathcal{R} + C_+ = \mathcal{R}$, on voit facilement que $\bigcup_{a \in \mathcal{R}} H(\mathcal{J}_+ \cup \mathcal{J}_- \cup (a+C_+))$ est un domaine d'holomorphie : c'est en effet le complémentaire de l'union des (n-1) - plans analytiques à coefficients réels dont la trace réelle ne rencontre pas $\mathcal{R}$ (utiliser Hahn Banach et la description du domaine par les droites réelles).

D. Généralisation à des Cônes convexes quelconques.

Soit Γ un cône convexe ouvert dans $\mathbb{R}^n$, posons $\mathcal{J}^\Gamma_\pm = \mathbb{R}^n \pm i\Gamma$. Soit $\mathcal{R}$ un ouvert de $\mathbb{R}^n$ tel que $\mathcal{R} + \Gamma = \mathcal{R}$. On peut toujours, par un changement de coordonnées, se ramener au cas où Γ contient C_+ . Il en résulte que 1°) $H(\mathcal{J}^\Gamma_+ \cup \mathcal{J}^\Gamma_- \cup \mathcal{R}) = H(\mathcal{J}^\Gamma_+ \cup \mathcal{J}^\Gamma_- \cup \text{conv.}\mathcal{R})$; 2°) $H(\mathcal{J}^\Gamma_+ \cup \mathcal{J}^\Gamma_- \cup \text{conv.}\mathcal{R}) = H(\mathcal{J}_+ \cup \mathcal{J}_- \cup \text{conv.}\mathcal{R})$ = union des points non réels des droites réelles qui rencontrent conv.$\mathcal{R}$ et de conv.$\mathcal{R}$ = complémentaire des (n-1)-plans analytiques réels qui ne coupent pas $\mathcal{R}$.

Finalement remarquons que si $\mathcal{R}$ est un ouvert réel (non nécessairement tel que $\mathcal{R}+\Gamma=\mathcal{R}$) contenant un cône ouvert convexe Γ_1 tel que $\Gamma\subset\Gamma_1$ on a

$$H((\mathbb{R}^n+i\Gamma)\cup(\mathbb{R}^n-i\Gamma)\cup\mathcal{R})\supset H((\mathbb{R}^n+i\Gamma)\cup(\mathbb{R}^n-i\Gamma)\cup\Gamma_1)=$$

$$=H((\mathbb{R}^n+i\Gamma_1)\cup(\mathbb{R}^n+i\Gamma_1)\cup\Gamma_1)$$

et par suite $H((\mathbb{R}^n+i\Gamma)\cup(\mathbb{R}^n-i\Gamma)\cup\mathcal{R})=H((\mathbb{R}^n+i\Gamma_1)\cup(\mathbb{R}^n-i\Gamma_1)\cup\mathcal{R})$.
Tous ces phénomènes ont des versions locales (visibles par transformations conformes par ex.) mais leur description n'est pas aussi simple.

E. Retour sur le cas où $\mathcal{R}$ est un double-cône.

Revenons au domaine $H(\mathcal{J}_+\cup\mathcal{J}_-\cup C(a,b))$, (n° A), avec

$\mathcal{J}_+=-\mathcal{J}_-=\{x+iy\in\mathbb{C}^n : y_j>0 \text{ pour tout } j\}$

$C(a,b)=\{x\in\mathbb{R}^n : a_i<x_i<b_j \text{ pour tout } j\}$.

Le changement de variables :

$$z'_j=1-\frac{b_j-a_j}{z_j-a_j}=\frac{z_j-b_j}{z_j-a_j}\quad,\quad(1\le j\le n)\quad,$$

applique $\mathcal{J}_\pm$ dans lui même et la région réelle $a+C_+$ sur $\{x\in\mathbb{R}^n : x_j<1\ (1\le j\le n)\}$. Il applique en particulier $C(a,b)$ sur C_- . On en déduit la description suivante du domaine

$H(\mathfrak{J}_+ \cup \mathfrak{J}_- \cup C(a,b)) =$

$$= \{z \in \mathbb{C}^n : \frac{z_j - b_j}{z_j - a_j} = t\lambda_j , \; \mathrm{Im}\lambda_j > 0 \; , \; \mathrm{Im}t > 0 \; , \; 1 \le j \le n\} \; .$$

Les n équations $\frac{z_j - b_j}{z_j - a_j} = t\lambda_j$, $(1 \le j \le n)$ où les λ_j sont fixés et où t décrit $\mathbb{C}$ sont les équations paramétriques d'une courbe algébrique. Ces courbes seront appelées "courbes Q" et jouent un rôle important dans les problèmes du type $H(\mathfrak{J}_+ \cup \mathfrak{J}_- \cup C(a,b))$. Elles passent par a et b .

F. Régions de coïncidence convexes.

On peut montrer que, dans le cas où $n = 2$, si la région de coïncidence est connexe et naturelle, on a

$$H(\mathfrak{J}_+ \cup \mathfrak{J}_- \cup \mathcal{R}) = \bigcup_{\substack{a \text{ et } b \in \mathcal{R} \\ b - a \in C_+}} W(a,b) \; .$$

(C'est un résultat de la représentation JLD, mais on peut également réobtenir JLD de cette façon). Ce résultat n'est plus vrai pour $n > 2$. Mais il reste valable si $\mathcal{R}$ est convexe. Le principe de la démonstration est le suivant. On suppose d'abord $\mathcal{R}$ bornée, convexe, et donnée par

$$\mathcal{R} = \{x : f(x) < 0 \; , \; g(x) > 0\} \quad ,$$

où f (resp g) est une fonction C^∞ telle que $\mathrm{grad} f$ (resp $\mathrm{grad} g$) $\in C_+$; on suppose que

$$\mathcal{R}_+ = \{x : g(x) > 0\} \quad \text{et} \quad \mathcal{R}_- = \{x : f(x) < 0\} \quad ,$$

sont convexes et telles que $\mathcal{R}_+ + C_+ = \mathcal{R}_+$, $\mathcal{R}_- + C_- = \mathcal{R}_-$.

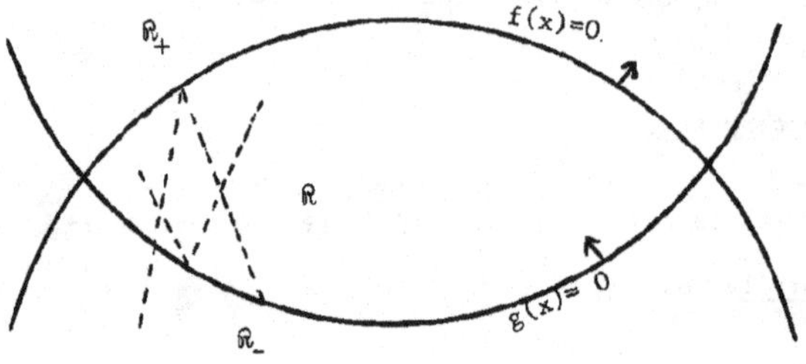

Figure 3

(Le passage au cas général, par passage à la limite, est facile).

Notons

$$W(\mathcal{R}) = \bigcup_{a,b \in \mathcal{R}} W(a,b) \quad .$$

L'étude, très longue et fastidieuse, de la frontière de $W(\mathcal{R})$, conduit aux résultats suivants :

1) la frontière est composée de points de $\partial\mathcal{T}_+ \cup \partial\mathcal{T}_-$ et d'un ensemble noté δW . Celui-ci comprend une partie "générique" $\delta_1 W$ et des parties dégénérées dont, en fait, les points sont adhérents à $\delta_1 W$.

2) $\delta_1 W$ est une union de "courbes Q" à coefficients réels. Tout point $z = (z_1,\dots,z_n)$ non réel de $\delta_1 W$ vérifie :

$$\frac{z_j - \xi_j}{z_j - \eta_j} = \lambda_j t$$

où : ξ et η sont des points réels tels que

$$f(\xi) = g(\eta) = 0 \quad , \quad \xi_j - \eta_j > 0 \quad \forall_j \quad ,$$

$$N = \operatorname{grad} f(\xi) \quad , \quad n = \operatorname{grad} g(\eta) \quad ,$$

$$\lambda_j = \epsilon_j \sqrt{\frac{n_j}{N_j}} \quad , \quad (\epsilon_j = \pm 1) \quad ,$$

$$\sum_{j=1}^{n} \omega_j(\xi_j - \eta_j) = 0 \quad \text{où} \quad \omega_j = \epsilon_j \sqrt{N_j\, n_j} \quad \text{pour tout } j \; ;$$

$$t \in \mathbb{C} \quad \text{et} \quad \operatorname{Im} t > 0 \quad .$$

Le calcul d'enveloppe qui conduit à ces conditions montre qu'en un tel point z , l'hyperplan tangent au domaine est donné par

$$\{\zeta : \sum_{j=1}^{n} \operatorname{Im} a_j(\zeta_j - z_j) = 0 \}$$

où

$$a_k = \frac{\bar{t}}{\operatorname{Im} t} P_k \quad , \quad P_k = N_k(1 - \lambda_k t)^2 = N_k \frac{(\xi_k - \eta_k)^2}{(z_k - \eta_k)^2} \quad .$$

En particulier cet hyperplan contient <u>l'hyperplan analytique</u> :

$$\{\zeta : \sum_{j=1}^{n} P_j(\zeta_j - z_j) = 0\} \quad .$$

On peut chercher l'enveloppe de ce dernier lorsque t varie. On trouve la variété d'équation :

$$\sum_{i,j} (N_i n_j - \omega_i \omega_j)(\zeta_i - \xi_i)(\zeta_j - \eta_j) = 0 \quad ,$$

ou (avec un produit scalaire évident) :

$$(N, \zeta - \xi)(n, \zeta - \eta) - (\omega, \zeta - \xi)(\omega, \zeta - \eta) = 0 \quad ,$$

ou encore, puisque $(\omega, \xi - \eta) = 0$:

$$(N, \zeta - \xi)(n, \zeta - \eta) - (\omega, \zeta - \xi)^2 = 0 \quad .$$

Pour tout x réel dans $\mathcal{R}$ on a (par convexité)

$$(N, x - \xi) < 0 \quad , \quad (n, x - \eta) > 0 \quad ,$$

tandis que si x appartient à la quadrique précédente, on a

$$(N, x - \xi)(n, x - \eta) = (\omega, x - \xi)^2 \geq 0 \quad .$$

La trace réelle de la quadrique ne coupe donc pas $\mathcal{R}$. On vérifie facilement que la quadrique (complexe) ne coupe pas $\mathcal{T}_+ \cup \mathcal{T}_-$.

Elle ne peut donc couper $H(\mathcal{T}_+ \cup \mathcal{T}_- \cup \mathcal{R})$. Ceci, joint à l'étude des autres points de la frontière, montre que

$$1°) \quad H(\mathcal{T}_+ \cup \mathcal{T}_- \cup \mathcal{R}) = W(\mathcal{R})$$
$$2°) \quad H(\mathcal{T}_+ \cup \mathcal{T}_- \cup \mathcal{R}) = \bigcap \bigcup_{A \in K} \{z : \sum_{|\alpha| \leq 2} A_\alpha z^\alpha = 0\}$$

où K est un ensemble de familles $A = \{A_\alpha\}_{|\alpha| \leq 2}$ de coefficients réels tels que la forme homogénéisée du polynôme quadratique $\sum_\alpha A_\alpha z^\alpha$ ait au plus 3 carrés, et que ce polynôme ne prenne aucune valeur positive dans le domaine.

G. <u>Régions convexes et cônes quelconques</u>.

Le résultat de (F) peut s'étendre au cas de tubes $\mathcal{T}_\pm^\Gamma = \mathbb{R}^n \pm i\Gamma$ où Γ est un cône ouvert convexe quelconque par la "méthode des variables surabondantes" : choisissons des vecteurs $e_1, \ldots, e_N$ dans Γ. Si f est analytique dans $H(\mathcal{T}_+^\Gamma \cup \mathcal{T}_-^\Gamma \cup \mathcal{R})$, la fonction

$$F(\zeta_1, \ldots, \zeta_N) = f\Big(\sum_{j=1}^{N} \zeta_j e_j\Big) ,$$

est en particulier, analytique dans l'union des deux tubes

$$\{\zeta \in \mathbb{C}^N : \operatorname{Im} \zeta_j > 0\} \text{ et } \{\zeta \in \mathbb{C}^N : \operatorname{Im} \zeta_j < 0\} ,$$

et au voisinage de

$$\dot{\mathcal{R}} = \{\xi \in \mathbb{R}^N : \sum_{j=1}^{N} \xi_j e_j \in \mathcal{R}\} .$$

Si $\mathcal{R}$ est convexe, $\mathring{\mathcal{R}}$ l'est aussi ; on peut lui appliquer la théorie de (F) et, au prix de quelques efforts supplémentaires, on obtient finalement le théorème suivant :

THEOREME 3.

Soit Γ un cône ouvert convexe de $\mathbb{R}^n$ tel que $\Gamma \cap (-\Gamma) = \emptyset$. Soit $\mathcal{R}$ un ouvert convexe de $\mathbb{R}^n$ tel que si $a \in \mathcal{R}$, $b \in \mathcal{R}$, $x-a \in \Gamma$, $b-x \in \Gamma$, alors $x \in \mathcal{R}$. Notons $W(a,b) = H(\mathcal{T}_+^\Gamma \cup \mathcal{T}_-^\Gamma \cup \Gamma(a,b))$, $\mathcal{T}_\pm^\Gamma = \mathbb{R}^n \pm i\Gamma$, $\Gamma(a,b) = \{x \in \mathbb{R}^n : x - a \in \Gamma , b - x \in \Gamma\}$. Alors

1°) $H(\mathcal{T}_+^\Gamma \cup \mathcal{T}_-^\Gamma \cup \mathcal{R}) = \bigcup_{a,b \in \mathcal{R}} W(a,b)$

2°) $H(\mathcal{T}_+^\Gamma \cup \mathcal{T}_-^\Gamma \cup \mathcal{R})$ est le complémentaire de l'union d'une famille de quadriques à coefficients réels ayant au plus 3 carrés (en coordonnées homogènes).

H. Cônes relativistes.

On considère $\mathbb{R}^{4n}$ (resp. $\mathbb{C}^{4n}$) comme produit topologique de n fois l'espace de Minkowski réel (resp.complexe). On notera si $z \in \mathbb{C}^{4n}$,

$$z = (z_1,\ldots,z_n) \text{ , } z_j \in \mathbb{C}^4 \text{ , } z_j = \{z_j^\mu\}_{\mu=0,1,2,3} = (z_j^0, \vec{z}_j)$$

$$(z_j,z_k) = z_j^\mu z_{k\mu} = z_j^\mu g_{\mu\nu} z_k^\nu = z_j^0 z_k^0 - z_j^1 z_k^1 - z_j^2 z_k^2 - z_j^3 z_k^3 .$$

Dans l'espace de Minkowski $\mathbb{R}^4$, le cône $V^+ = -V^-$ est défini par

$$V^+ = \{x = \{x^0 , \vec{x}\} : x^0 > |\vec{x}| \} .$$

On pose $\mathcal{T}^+ = \mathbb{R}^4 + iV^+$. Parmi les automorphismes analytiques de $\mathcal{T}^+$ figure l'inversion relativiste :

$$z = \{z^\mu\} \rightarrow - \frac{z}{(z,z)} ,$$

(noter que (z,z) ne s'annule pas dans $\mathcal{T}^+$) . On en profite pour calculer simplement le domaine e.o.w. $H((\mathcal{T}^+)^n \cup (\mathcal{T}^-)^n \cup \Gamma(a,b))$ où

$$\Gamma(a,b) = \{x \in \mathbb{R}^{4n} : x - a \in (V^+)^n , - (x - b) \in (V^+)^n\} .$$

En effet la transformation

$$z_j \rightarrow z_j' = - \frac{z_j - a_j}{(z_j-a_j,z_j-a_j)} \qquad (j = 1,\ldots,n) ,$$

conserve les tubes $(\mathcal{T}^\pm)^n$ et applique $\Gamma(a,b)$ sur

$$\{x = (x_1,\ldots,x_n) \in \mathbb{R}^{4n} : \forall_j , x_j + \frac{b_j - a_j}{(b_j-a_j,b_j-a_j)} \in V^-\} .$$

D'après la théorie générale (voir n° D) l'enveloppe $H((\mathcal{T}^+)^n \cup (\mathcal{T}^+)^n \cup c - (V^+)^n)$ est donnée par

$$\{z' : z_j' - c_j = t\lambda_j , \mathrm{Im} t > 0 , \mathrm{Im}\lambda_j \in V^+ (1 \leq j \leq n)\}$$

donc

$$H((\mathcal{T}^+)^n \cup (\mathcal{T}^-)^n \cup \Gamma(a,b)) =$$

$$= \{z : \forall_j , z_j - a_j = - \frac{t\lambda_j - c_j}{(t\lambda_j-c_j,t\lambda_j-c_j)} , \mathrm{Im} t > 0, \mathrm{Im}\lambda_j \in V^+\}$$

où $$c_j = - \frac{(b_j - a_j)}{(b_j-a_j,b_j-a_j)} .$$

De plus le complémentaire du domaine est l'union des inverses des hyperplans analytiques à coefficients réels qui ne rencontrent pas $c+(V^-)^n$ dans les réels. En particulier, si $n = 1$ on obtient des hyperboloïdes : l'inverse du plan $\{z' : (h,z') + k = 0\}$ s'écrit en effet

$$-\frac{(h, z-a)}{(z-a,z-a)} + k = 0 \quad \text{, ou encore :}$$

$$k(z-a, z-a) - (h, z-a) = 0 \quad .$$

C'est un hyperboloïde (si $k \neq 0$) ayant pour cône asymptotique le cône de lumière $\{z : (z,z) = 0\}$ et passant par a . En outre le domaine a pour points non réels tous les points non réels appartenant à des inverses de droites réelles (c'est-à-dire des hyperboles réelles asymptotiques à des directions du genre lumière) dont la trace réelle rencontre $\Gamma(a,b)$. Ces remarques permettent, en fait, de retrouver le résultat de Jost-Lehmann-Dyson.

II.2.- E.O.W. OBLIQUE.

Dans ce sous-paragraphe, on notera C_1 et C_2 deux cônes non vides convexes dans $\mathbb{R}^n$ tels que $C_1 \cap (-C_2) = \emptyset$, $C_+ = C_1 - C_2 = -C_-$, $\mathcal{J}_1 = \mathbb{R}^n + iC_1$, $\mathcal{J}_2 = \mathbb{R}^n + iC_2$, $\mathcal{J}_\pm = \mathbb{R}^n \pm iC_+$.

L'exemple traité au §I. est un problème d'e.o.w. oblique. La solution peut être interprétée comme une intersection :

$$H(\{(z_1,z_2) \in \mathbb{C}^2 : \operatorname{Im} z_2 = 0 , 0 < \operatorname{Im} z_1 < \pi\} \cup \{z_1,z_2 \in \mathbb{C}^2 : \operatorname{Im} z_1 = 0, 0 < \operatorname{Im} z_2 < \pi\} \cup \mathbb{R}^2) =$$

$$= H(\{(z_1,z_2) \in \mathbb{C}^2 : 0 < \operatorname{Im} z_1 < \pi, 0 < -\operatorname{Im} z_2 < \pi\} \cup \{z_1,z_2 \in \mathbb{C}^2 : 0 < -\operatorname{Im} z_1 < \pi ,$$

$$0 < \operatorname{Im} z_2 < \pi\} \cup \mathbb{R}^2) \cap \{z_1,z_2 \in \mathbb{C}^2 : 0 < \operatorname{Im} z_1 < \pi, 0 < \operatorname{Im} z_2 < \pi\} .$$

En faisant les changements de variables indiqués au §I. on en déduit que

$$H(\{(z_1,z_2) \in \mathbb{C}^2 : \operatorname{Im} z_2 = 0, -\tau < z_2 < \tau , \operatorname{Im} z_1 > 0\} \cup$$

$$\cup \{(z_1,z_2) \in \mathbb{C}^2 : \operatorname{Im} z_1 = 0, -\tau < z_1 < \tau , \operatorname{Im} z_2 > 0\}$$

$$\cup \{(x_1,x_2) \in \mathbb{R}^2 : -\tau < x_1 < \tau , -\tau < x_2 < \tau\} =$$

$$= H(\{(z_1,z_2) \in \mathbb{C}^2 : \operatorname{Im} z_1 > 0 , \operatorname{Im} z_2 < 0\} \cup \{(z_1,z_2) \in \mathbb{C}^2 : \operatorname{Im} z_1 < 0 , \operatorname{Im} z_2 > 0 \}$$

$$\cup \{(x_1,x_2) \in \mathbb{R}^2 : |x_1| < \tau , |x_2| < \tau\} .$$

En d'autres termes si C_1 est le cône $\{(y_1,y_2) \in \mathbb{R}^2 : y_1 > 0 , y_2 = 0\}$, $C_2 = \{(y_1,y_2) \in \mathbb{R}^2 : y_2 > 0, y_1 = 0\}$ et $\mathcal{R}$ est le double-cône $\{x \in \mathbb{R}^2 : |x_1| < \tau , |x_2| < \tau \}$ on a

$$H(\mathcal{J}_1 \cup \mathcal{J}_2 \cup \mathcal{R}) = H(\mathcal{J}_+ \cup \mathcal{J}_- \cup \mathcal{R}) \cap (\mathcal{J}_1 + \mathcal{J}_2) .$$

Ici le "double-cône" doit être interprété comme fabriqué avec C_+ . Il est facile de constater que l'équation ci-dessus reste vraie, à deux dimensions, si C_1 et C_2 sont des cônes ouverts convexes quelconques tels que $C_1 \cap (-C_2) = \emptyset$ et si $\mathcal{R}$ est naturelle pour $\mathcal{J}^+$ i.e. si

$$H(\mathcal{J}_+ \cup \mathcal{J}_- \cup \mathcal{R}) \cap \mathbb{R}^2 = \mathcal{R} .$$

Ce théorème général peut d'ailleurs se démontrer directement en appliquant la solubilité du problème de Cousin dans les domaines d'holomorphie (voir [1]) . Nous avons vu au § I. un phénomène d'agrandissement de la région de coïncidence. Il est clair qu'à deux dimensions on obtient immédiatement le lemme suivant :

Lemme 4.

Soient C_1 et C_2 des cônes convexes dans $\mathbb{R}^2$, non réduits à $\{0\}$, tels que $C_1 \cap (-C_2) = \emptyset$; si $\mathcal{R}$ est un ouvert réel de $\mathbb{R}^2$ contenant un segment de droite d'extrémités a et b telles que $b - a \in C_+$ $(= C_1 - C_2)$ alors $\tilde{\mathcal{R}} = H(\mathcal{J}_1 \cup \mathcal{J}_2 \cup \mathcal{R}) \cap \mathbb{R}^2$ contient le double cône $(a + C_+) \cap (b - C_+)$.

Montrons maintenant que le deuxième phénomène d'agrandissement de la région de coïncidence (voir II.1.C) se produit aussi. Considérons à nouveau le cas $C_1 = \{y_1, y_2 : 0 < y_1, y_2 = 0\}$, $C_2 = \{y_1, y_2 : 0 < y_2, y_1 = 0\}$ et prenons $\mathcal{R} = C_+ = \{y_1, y_2 : y_1 > 0, y_2 < 0\}$. Le théorème d'intersection s'applique à ce cas (cas limite d'un double-cône) donc

$$H(\mathcal{J}_1 \cup \mathcal{J}_2 \cup \mathcal{R}) = H(\mathcal{J}_+ \cup \mathcal{J}_- \cup \mathcal{R}) \cap (\mathcal{J}_1 + \mathcal{J}_2) =$$

$$= H(\mathcal{J}_+ \cup \mathcal{J}_- \cup \mathcal{R}) \cap \{z_1, z_2 : \mathrm{Im} z_1 > 0 , \mathrm{Im} z_2 > 0\} .$$

En particulier les points non réels du domaine sont les points non réels de toute droite réelle dont la trace réelle rencontre $\mathcal{R}(= C_+)$ <u>et dont la partie imaginaire est contenue dans</u> $C_1 + C_2$. Supposons maintenant que $\mathcal{R}$ est une union de translatés de C_+ . Soit $\tilde{\mathcal{R}}$ = Conv. $\mathcal{R}$. Si $x \in \tilde{\mathcal{R}}$, toute droite réelle passant par x rencontre $\mathcal{R}$. Il en résulte que $H(\mathcal{J}_1 \cup \mathcal{J}_2 \cup \mathcal{R})$ contient l'intersection d'un voisinage complexe de x avec $\mathcal{J}_1 + \mathcal{J}_2$ et que la région de coïncidence s'étend jusqu'à $\tilde{\mathcal{R}}$. Il est évident qu'à deux dimensions, ce résultat s'étend à des C_1 , C_2 convexes non vides quelconques tels que $C_1 \cap (-C_2) = \emptyset$.

Pour $n > 2$ le théorème d'intersection n'est plus vrai en général. Toutefois on a le théorème suivant.

<u>THEOREME 5</u>.

Soient C_1 et C_2 deux cônes ouverts convexes non vides de $\mathbb{R}^n$ tels que $C_1 \cap (-C_2) = \emptyset$; on pose $C_+ = C_1 - C_2$, $\mathcal{J}_1 = \mathbb{R}^n + iC_1$, $\mathcal{J}_2 = \mathbb{R}^n + iC_2$, $\mathcal{J}_\pm = \mathbb{R}^n \pm iC_+$. Soit $\mathcal{R}$ un ouvert convexe tel que $\mathcal{R} + C_+ = \mathcal{R}$. Alors

$$H(\mathcal{J}_1 \cup \mathcal{J}_2 \cup \mathcal{R}) = H(\mathcal{J}_+ \cup \mathcal{J}_- \cup \mathcal{R}) \cap (\mathcal{J}_1 + \mathcal{J}_2) .$$

<u>Démonstration</u> : Elle se fait par la méthode des variables surabondantes et ne sera pas donnée en détail.

D'autres exemples de cas où le théorème d'intersection est vrai se rencontrent en théorie des champs : voir par ex. [3] et [4] .

Nous terminerons ces remarques sur l'e.o.w. oblique par le théorème suivant (cf. [2] . Pour le cas de l'e.o.w. opposé, le théorème est dû à Dyson [7] , Borchers [8] , Vladimirov [9]).

THEOREME 6.

Soient C_1 et C_2 deux cônes ouverts convexes non vides de $\mathbb{R}^n$, $C_+ = C_1 - C_2$, $\mathcal{J}_1 = \mathbb{R}^n + iC_1$, $\mathcal{J}_2 = \mathbb{R}^n + iC_2$. Soient a et b deux points de $\mathbb{R}^n$ tels que $b - a \in C_+$, V un voisinage ouvert réel du segment (ouvert) (a, b), et W un voisinage ouvert complexe du "double-cône" :

$$K = \{x \in \mathbb{R}^n \; : \; x - a \in C_+ \, , \, b - x \in C_+\} \, .$$

Si f_1 et f_2 sont des fonctions holomorphes dans $\mathcal{J}_1 \cap W$ et $\mathcal{J}_2 \cap W$, respectivement, et ont des valeurs aux bords dans K qui coïncident dans V, elles coïncident dans tout le double-cône K.

Remarque : La condition que C_1 et C_2 soient ouverts est essentielle dans le cas $n \geq 3$: on connaît des contre-exemples lorsqu'elle n'est pas satisfaite, même si C_+ est ouvert.

Démonstration :

a) Considérons d'abord un cas particulier. Soient p et q deux entiers > 1 ; un point courant de $\mathbb{C}^p \times \mathbb{C}^q$ sera noté (z,w) avec $z = x + iy$, $w = u + iv$, $x = (x_1,\dots,x_p)$ et $y = (y_1,\dots,y_p) \in \mathbb{R}^p$, $u = (u_1,\dots,u_q)$ et $v = (v_1,\dots,v_q) \in \mathbb{R}^q$. On suppose que f_1 est holomorphe dans l'ouvert D_1 de $\mathbb{C}^p \times \mathbb{C}^q$ donné par

$$D_1 = \{(z,w) \; : \; z_j \in D(\tau,\alpha) \text{ pour tout } j = 1,\dots,p \; ; \; |u_k| < \tau \text{ et } v_k < \epsilon y_1 \text{ pour tout } k = 1,\dots,q\}$$

et f_1 est holomorphe dans

$$D_2 = \{(z,w) \; : \; w_k \in D(\tau,\alpha) \text{ pour tout } k = 1,\dots,q \; ; \; |x_j| < \tau \text{ et } y_j < \epsilon v_1 \text{ pour tout } j = 1,\dots,p\} \, .$$

f_j est C^∞ dans $\overline{D}_j$ $(j = 1,2)$ et les valeurs aux bords sur les réels de ces deux fonctions coïncident dans

$$V = \{(x,u) \in \mathbb{R}^p \times \mathbb{R}^q \ : \ |x_j - x_1| < \delta \ , \ |x_j| < \tau \ \text{pour tout} \ j \ ,$$

$$|u_k + x_1| < \delta \ , \ |u_k| < \tau \ \text{pour tout} \ k\} \ .$$

La conclusion est que f_1 et f_2 coïncident dans tout le double-cône $\{(x,u) \in \mathbb{R}^p \times \mathbb{R}^q \ : \ |x_j| < \tau \ , \ |u_k| < \tau \ , \ \forall j \ , \ \forall k \}$.

Pour le voir posons

$$z_j = \tau \operatorname{th} \frac{\alpha \zeta_j}{2} \qquad , \ (\zeta_j = \xi_j + i\eta_j = \frac{1}{\alpha} \log \frac{\tau + z_j}{\tau - z_j})$$

$$w_k = \tau \operatorname{th} \frac{\alpha \lambda_k}{2} \qquad , \ (\lambda_k = \rho_k + i\sigma_k = \frac{1}{\alpha} \log \frac{\tau + w_k}{\tau - w_k}) \ ;$$

et $$g_j(\xi,w) = \frac{f_j(\tau \operatorname{th} \frac{\alpha \zeta_1}{2}, \ldots, \tau \operatorname{th} \frac{\alpha\lambda_q}{2})}{\prod_{r=1}^{p} (\zeta_r + i) \prod_{k=1}^{q} (\lambda_k + i)} \ , \quad j = 1,2 \ .$$

On vérifie facilement que les domaines d'analyticité de g_1 et g_2 contiennent respectivement les domaines D''_1 et D''_2 donnés par

$$D''_1 = \{\zeta, \lambda : 0 < \eta_j < 1 \ , \ 0 < \sigma_k < 1 \ ,$$

$$0 < e^{-\alpha|\rho_k|} \sigma_k < \epsilon' \ \eta_1 \ e^{-\alpha|\xi_1|} \} \ ,$$

$$D''_2 = \{\zeta, \lambda : 0 < \eta_j < 1 \ , \ 0 < \sigma_k < 1 \ ,$$

$$0 < e^{-\alpha|\xi_j|} \eta_j < \epsilon' \ \sigma_1 \ e^{-\alpha|\rho_1|} \} \ ,$$

(où $\varepsilon' = \frac{\varepsilon}{4} \cos^2 \frac{\alpha}{2}$) . De plus g_1 et g_2 coïncident aux points réels (ξ,ρ) de V'

$$V' = \{(\xi,\rho) : |\xi_j - \xi_1| < \frac{\delta}{\alpha\tau} , \ |\rho_k + \xi_1| < \frac{\delta}{\alpha\tau} \ (j=1,..,p, \ k=1,..,q)\} .$$

Posons

$$h(\zeta,\lambda) = \frac{1}{2\pi i} \int_{-\infty}^{\infty} \frac{dt}{t-\lambda_1} \ g_1(\zeta_1+\lambda_1-t,\ldots,\zeta_p+\lambda_1-t,t,\lambda_2-\lambda_1+t,..,\lambda_q-\lambda_1+t)$$

$$+\frac{1}{2\pi i} \int_{-\infty}^{\infty} \frac{ds}{s-\zeta_1} \ g_2(s,\zeta_2-\zeta_1+s,\ldots,\zeta_p-\zeta_1+s,\lambda_2+\zeta_1-s,..,\lambda_q+\zeta_1-s) .$$

Il est facile de vérifier que h est holomorphe dans

$$D_3'' = \{\zeta,\lambda : 0 < \eta_j < \tfrac{1}{2} , \ 0 < \sigma_k < \tfrac{1}{2} , \ 0 < \sigma_k - \sigma_1 < \varepsilon'(\eta_1+\sigma_1)e^{-4M(\xi,\rho)}$$

$$0 < \eta_j - \eta_1 < \varepsilon'(\eta_1+\sigma_1)e^{-4M(\xi,\rho)}, \ \forall j = 1,..,p \ , \ \forall k = 1,..,q\} ,$$

où $M(\xi,\rho) = \max (|\xi_j| , |\rho_k|)$.

[Toutes ces estimations reposent sur les inégalités suivantes : si

$$y = \operatorname{Im} \operatorname{th} \frac{\alpha}{2} (\xi + i\eta) \ , \ 0 < \alpha < \pi \ , \ 0 < \eta < 1 \ ,$$

on a :

$$\tfrac{1}{2}\eta e^{-|\alpha\xi|} \sin \alpha < \tfrac{1}{2} \sin \alpha\eta e^{-|\alpha\xi|} < y < 4e^{-\alpha|\xi|} \operatorname{tg}\frac{\alpha\eta}{2} < 4\eta e^{-\alpha|\xi|} \operatorname{tg} \frac{\alpha}{2}] .$$

Les domaines D_3'' et D_1'' (resp. D_3'' et D_2'') ont une intersection non vide qui contient, pour tout (ξ,ρ) , l'intersection d'un voisinage complexe de (ξ,ρ) avec un tube à base conique. De plus, par un calcul tout à fait analogue à celui du §I., on constate que

$h(\xi,\rho) = g_1(\xi,\rho) = g_2(\xi,\rho)$ dans le domaine V' . Donc $h(\xi,\rho)$, $g_1(\xi,\rho)$ et $g_2(\xi,\rho)$ coïncident partout et notre assertion est vérifiée.

b) Cas général.

On se place dans les hypothèses du théorème 5 en supposant que $a = 0$, et b est fixé dans C_+ . Choisissons des vecteurs $e_1,\dots, e_p \in C_1$, $e_{p+1},\dots, e_N \in - C_2$ de telle sorte que $b = \sum_{j=1}^{N} e_j$. On note $\Gamma_1 = \{x = \sum_{j=1}^{p} \xi_j e_j \ , \ \xi_j > 0\}$, $\Gamma_2 = \{x = \sum_{j=p+1}^{N} \xi_j e_j \ , \ \xi_j < 0\}$, et $\Gamma_+ = \Gamma_1 - \Gamma_2$. Le double cône K est donné par

$$\{x \in \mathbb{R}^n : x \in C_+ \ , \ b-x \in C_+\} \ .$$

Définissons deux nouvelles fonctions analytiques dans des domaines de $\mathbb{C}^N$ par :

$$F_j(\zeta) = f_j\left(\sum_{k=1}^{N} \zeta_k e_k\right) \ .$$

En supposant qu'on a légèrement réduit les domaines d'analyticité de f_1 et f_2 on se ramène au cas où F_1 (resp. F_2) est analytique dans l'intersection de l'ouvert :

$$\{\zeta \in \mathbb{C}^N : \left|\operatorname{Arg} \frac{\zeta_j}{1-\zeta_j}\right| < \alpha \ , \ 1 \le j \le N\} \ ,$$

avec un tube à base cônique convexe ouverte contenant tous les points non réels de :

$$\{\zeta \in \mathbb{C}^N : \operatorname{Im} \zeta_j \ge 0 \text{ pour } 1 \le j \le p \ , \ \operatorname{Im} \zeta_j = 0 \text{ pour } p+1 \le j \le N\}$$

$$(\text{resp. } \{\zeta \in \mathbb{C}^N : \operatorname{Im} \zeta_j = 0 \text{ pour } 1 \le j \le p \ , \ \operatorname{Im} \zeta_j \le 0 \text{ pour } p+1 \le j \le N\}) \ .$$

De plus les valeurs aux bords de F_1 et F_2 aux points réels coïncident dans un voisinage ouvert réel de tout segment réel de la forme

$$\{\xi \in \mathbb{R}^N : \xi_j = \theta(1+\tau_j) \ , \ 0 \leq \theta \leq 1\} \ ,$$

où les τ_j sont des nombres réels tels que $1+\tau_j > 0$ pour tout j et $\sum_{j=1}^{N} \tau_j e_j = 0$. Soit U l'ensemble des $\tau \in \mathbb{R}^N$ possédent ces propriétés.
En appliquant le cas (a) on voit que F_1 et F_2 coïncident dans l'union lorsque τ parcourt U , des ouverts réels

$$\{\xi \in \mathbb{R}^N : \xi_j = \theta_j(1+\tau_j) \ , \ 0 < \theta_j < 1\} \ ,$$

et que f_1 et f_2 coïncident donc dans l'union, lorsque τ parcourt U , des ouverts réels

$$\{x \in \mathbb{R}^n : x = \sum_{j=1}^{N} \theta_j(1+\tau_j)\, e_j \ , \ 0 < \theta_j < 1\} \ .$$

Cette union n'est autre que le double cône

$$\{x \in \mathbb{R}^n : x \in \Gamma^+, \ (b-x) \in \Gamma^+\} \ , \text{ où } \Gamma^+ = \Gamma_1 - \Gamma_2,$$

$$\Gamma^+ = \{x \in \mathbb{R}^n : x = \sum_{i=1}^{N} \xi_i e_i \ , \ \xi_i > 0 \ \forall i\} \ .$$

En effet si $x = \sum_{j=1}^{N} \theta_j(1+\tau_j)\, e_j$ avec $1+\tau_j > 0$, $0 < \theta_j < 1$, et $\sum_j \tau_j e_j = 0$, on a évidemment $x = \sum_j \xi_j e_j$ avec $\xi_j > 0$ donc $x \in \Gamma^+$ et d'autre part $\sum_j e_j - x = \sum_{j=1}^{N} \{(1-\theta_j) - \theta_j \tau_j\} e_j = \sum_{j=1}^{N} (1-\theta_j)\tau_j e_j$ donc $b-x \in \Gamma^+$. Réciproquement si $x \in \Gamma^+$ et $b-x \in \Gamma^+$ on a

$$x = \sum_{j=1}^{N} \xi_j e_j = \sum_{j=1}^{N} (1-u_j)\, e_j \quad \text{où } \xi_j > 0 \ , \ u_j > 0 \ .$$

Donc $\sum_{j=1}^{N} (\xi_j + u_j - 1)\, e_j = 0$. Posons $\tau_j = \xi_j + u_j - 1$, et $\theta_j = \dfrac{\xi_j}{\xi_j + u_j}$; on a $x = \sum_j \theta_j (1+\tau_j) e_j$.

En faisant tendre Γ_1 et Γ_2 vers C_1 et C_2 on obtient le théorème.

III. - EXTENSIONS PAR DES GROUPES ANALYTIQUES.

Ce dernier paragraphe n'a que des rapports lointains avec les questions de valeurs aux bords etc... ; il figure ici parcequ'il présente un mécanisme, très important en théorie des champs, d'agrandissement des domaines. Commençons par un lemme très simple :

Lemme 7.

Soit D un domaine de $\mathbb{C}^n \times \mathbb{C}^m$ de la forme suivante :

$$D = \{(z,w) \in \mathbb{C}^n \times \mathbb{C}^m : z \in \Omega, \quad w \in \Delta(z)\} ,$$

où Ω est un domaine de $\mathbb{C}^n$ et $\Delta(z)$ est, pour chaque $z \in \Omega$, un domaine non vide de $\mathbb{C}^m$. S'il existe un ouvert Ω_1 dans Ω tel que $z \in \Omega_1 \Rightarrow \Delta(z) = \mathbb{C}^m$ (c'est à dire si D contient $\Omega_1 \times \mathbb{C}^m$) l'enveloppe d'holomorphie de D contient $\Omega \times \mathbb{C}^m$.

Démonstration

a) un cas particulier

Supposons que $\Omega = \{z : |z_j + iR_j| < R_j , \ 1 \leq j \leq n\}$, $\Omega_1 = \{z : |z_j + ir_j| < r_j , \ 1 \leq j \leq n\}$, avec $r_j < R_j$, et :

$$D = (\Omega_1 \times \mathbb{C}^m) \cup (\Omega \times \{w : |w_k| < \tau_k\}) .$$

L'inversion $z_j = \frac{1}{z'_j}$ et la transformation $w_k = e^{-iw'_k}$ transforment ce domaine en un tube :

$$D' = \{z',w' : \operatorname{Im} z'_j > \frac{1}{2r_j}\} \cup \{z',w' : \operatorname{Im} z'_j > \frac{1}{2R_j} , \ \operatorname{Im} w'_k < \log \tau_k\} ,$$

dont l'enveloppe d'holomorphie est : $\{z',w' : \operatorname{Im} z'_j > \frac{1}{2R_j}\} = D''$. De plus toute fonction holomorphe dans D' et périodique en $\operatorname{Re} w'_k$ se prolonge dans D'' en une fonction périodique. Il en résulte que l'enveloppe d'holomorphie de D est bien $\Omega \times \mathbb{C}^m$.

b) Le cas général est obtenu en construisant dans Ω des chaînes finies de polydisques (s'intersectant 2 à 2) commençant dans Ω_1 et en utilisant le cas (a). Les détails sont laissés au lecteur.

THEOREME 8.

Soit G un groupe de Lie complexe, $\mathfrak{G}$ son algèbre de Lie et notons $X \to \exp X$ l'application exponentielle habituelle. Supposons donnée une application

$$\begin{gathered}(g,z) \mapsto g.z \\ G \times \mathbb{C}^n \to \mathbb{C}^n\end{gathered} ,$$

holomorphe et telle que $1.z = z$, $h.(g.z) = (hg).z$. Soit Ω un domaine de $\mathbb{C}^n$; supposons qu'il existe un ouvert $\Omega_1 \subset \Omega$ tel que $G.\Omega_1 \subset \Omega$. Alors l'enveloppe d'holomorphie de Ω est un domaine de Riemann dont la projection sur $\mathbb{C}^n$ contient $G.\Omega$. Plus précisément soit $\hat{G}$ le groupe de recouvrement de G , τ l'homomorphisme naturel de $\hat{G}$ sur G . Si f est holomorphe dans Ω , la fonction

$$(h,z) \longmapsto f(\tau(h).z) ,$$

a un prolongement unique F holomorphe sur le produit topologique $\hat{G} \times \Omega$.

Démonstration.

Considérons une suite finie $X_1,\dots,X_m$ d'éléments non nuls de $\mathcal{G}$ et posons

$$H_{X_1,\dots,X_m}(\zeta_1,\dots,\zeta_m;z) = f(\exp\zeta_1 X_1.\exp\zeta_2 X_2.\ \dots\exp\zeta_m X_m.z) \quad .$$

D'après le lemme 6 cette fonction a un prolongement analytique unique dans $\mathbb{C}^m \times \Omega$. Le théorème s'obtient par des raisonnements de monodromie faciles qui ne seront pas reproduits ici (voir [10]) .

En pratique on utilise plutôt une autre version du théorème 7 :

THEOREME 9.

Soit G un groupe de Lie complexe, $\mathcal{G}$ son algèbre de Lie, et $(g,z)\mapsto g.z$ une application holomorphe de $G\times\mathbb{C}^n$ dans $\mathbb{C}^n$ telle que $1.z = z$ et $h.(g.z) = (hg).z$. Soit $\mathfrak{F}$ une base de $\mathcal{G}$. Soit Ω un domaine de $\mathbb{C}^n$ tel que, pour tout $X\in\mathfrak{F}$, il existe un ouvert $V_X \subset \Omega$ tel que $(\exp \xi X).V_X \subset \Omega$ pour tout $\xi\in\mathbb{C}$. Alors on a la même conclusion que dans le théorème 8.

(Dans ce cas, pour démontrer l'analyticité de $H_{X_1,\dots,X_m}$, on montre d'abord que $H_{X_1}(\zeta_1\ ;\ z)$ est analytique dans $\mathbb{C}\times\Omega$. On considère alors $H_{X_1}(\zeta_1\ ;\ \exp\zeta_2 X_2.z) = H_{X_1,X_2}(\zeta_1,\zeta_2\ ;z)$ qui est analytique dans $\mathbb{C}\times\mathbb{C}\times\Omega$, etc... . Le reste de la démonstration est inchangé).

Ces théorèmes sont appliqués, en théorie des champs, au cas où G est le groupe de Lorentz complexe connexe. La version donnée ici suit et généralise légèrement celle de [10] . Voir [11] et [12] .

REFERENCES

[1] J. Bros. Les problèmes de construction d'enveloppe d'holomorphie en théorie quantique des champs, Séminaire Lelong 4ème année n° 8 , 1962 .

[2] H. Epstein dans : Axiomatic Field Theory, Chretien & Deser éditeurs, Gordon & Breach, New York 1966.

[3] J. Bros, H. Epstein et V. Glaser, Nuovo Cimento 31, 1965 (1964).

[4] H. Epstein, V. Glaser et A. Martin, Commun. Math. Phys. 13, 257 (1969).

[5] J. Bros, H. Epstein et V. Glaser, Helv. Phys. Acta .

[6] R. Jost et H. Lehmann, Nuovo Cimento 5, 1598 (1957).

[7] F.J. Dyson, Phys. Rev. 110, 1460 (1958).

[8] H.J. Borchers, Nuovo Cimento 19, 781 (1961).

[9] V.S. Vladimirov, Trudy Mat. Inst. A.N. SSSR, 60, 101 (1961).

[10] J. Bros, H. Epstein et V. Glaser, Commun. Math. Phys. 6, 77 (1967).

[11] R.F. Streater, J. Math. Phys. 3, 256 (1962) .

[12] R. Jost, General Theory of Quantized Fields, American Math. Soc. Providence, R.I. 1965.

Vol. 277: Séminaire Banach. Edité par C. Houzel. VII, 229 pages. 1972. DM 22,–

Vol. 278: H. Jacquet, Automorphic Forms on GL(2). Part II. XIII, 142 pages. 1972. DM 18,–

Vol. 279: R. Bott, S. Gitler and I. M. James, Lectures on Algebraic and Differential Topology. V, 174 pages. 1972. DM 20,–

Vol. 280: Conference on the Theory of Ordinary and Partial Differential Equations. Edited by W. N. Everitt and B. D. Sleeman. XV, 367 pages. 1972. DM 29,–

Vol. 281: Coherence in Categories. Edited by S. Mac Lane. VII, 235 pages. 1972. DM 22,–

Vol. 282: W. Klingenberg und P. Flaschel, Riemannsche Hilbertmannigfaltigkeiten. Periodische Geodätische. VII, 211 Seiten. 1972. DM 22,–

Vol. 283: L. Illusie, Complexe Cotangent et Déformations II. VII, 304 pages. 1972. DM 27,–

Vol. 284: P. A. Meyer, Martingales and Stochastic Integrals I. VI, 89 pages. 1972. DM 18,–

Vol. 285: P. de la Harpe, Classical Banach-Lie Algebras and Banach-Lie Groups of Operators in Hilbert Space. III, 160 pages. 1972. DM 18,–

Vol. 286: S. Murakami, On Automorphisms of Siegel Domains. V, 95 pages. 1972. DM 18,–

Vol. 287: Hyperfunctions and Pseudo-Differential Equations. Edited by H. Komatsu. VII, 529 pages. 1973. DM 40,–

Vol. 288: Groupes de Monodromie en Géométrie Algébrique. (SGA 7 I). Dirigé par A. Grothendieck. IX, 523 pages. 1972. DM 55,–

Vol. 289: B. Fuglede, Finely Harmonic Functions. III, 188. 1972. DM 20,–

Vol. 290: D. B. Zagier, Equivariant Pontrjagin Classes and Applications to Orbit Spaces. IX, 130 pages. 1972. DM 18,–

Vol. 291: P. Orlik, Seifert Manifolds. VIII, 155 pages. 1972. DM 18,–

Vol. 292: W. D. Wallis, A. P. Street and J. S. Wallis, Combinatorics: Room Squares, Sum-Free Sets, Hadamard Matrices. V, 508 pages. 1972. DM 55,–

Vol. 293: R. A. DeVore, The Approximation of Continuous Functions by Positive Linear Operators. VIII, 289 pages. 1972. DM 27,–

Vol. 294: Stability of Stochastic Dynamical Systems. Edited by R. F. Curtain. IX, 332 pages. 1972. DM 29,–

Vol. 295: C. Dellacherie, Ensembles Analytiques Capacités. Mesures de Hausdorff. XII, 123 pages. 1972. DM 18,–

Vol. 296: Probability and Information Theory II. Edited by M. Behara, K. Krickeberg and J. Wolfowitz. V, 223 pages. 1973. DM 22,–

Vol. 297: J. Garnett, Analytic Capacity and Measure. IV, 138 pages. 1972. DM 18,–

Vol. 298: Proceedings of the Second Conference on Compact Transformation Groups. Part 1. XIII, 453 pages. 1972. DM 35,–

Vol. 299: Proceedings of the Second Conference on Compact Transformation Groups. Part 2. XIV, 327 pages. 1972. DM 29,–

Vol. 300: P. Eymard, Moyennes Invariantes et Représentations Unitaires. II, 113 pages. 1972. DM 18,–

Vol. 301: F. Pittnauer, Vorlesungen über asymptotische Reihen. VI, 186 Seiten. 1972. DM 18,–

Vol. 302: M. Demazure, Lectures on p-Divisible Groups. V, 98 pages. 1972. DM 18,–

Vol. 303: Graph Theory and Applications. Edited by Y. Alavi, D. R. Lick and A. T. White. IX, 329 pages. 1972. DM 26,–

Vol. 304: A. K. Bousfield and D. M. Kan, Homotopy Limits, Completions and Localizations. V, 348 pages. 1972. DM 29,–

Vol. 305: Théorie des Topos et Cohomologie Etale des Schémas. Tome 3. (SGA 4). Dirigé par M. Artin, A. Grothendieck et J. L. Verdier. VI, 640 pages. 1973. DM 55,–

Vol. 306: H. Luckhardt, Extensional Gödel Functional Interpretation. VI, 161 pages. 1973. DM 20,–

Vol. 307: J. L. Bretagnolle, S. D. Chatterji et P.-A. Meyer, Ecole d'été de Probabilités: Processus Stochastiques. VI, 198 pages. 1973. DM 22,–

Vol. 308: D. Knutson, λ-Rings and the Representation Theory of the Symmetric Group. IV, 203 pages. 1973. DM 22,–

Vol. 309: D. H. Sattinger, Topics in Stability and Bifurcation Theory. VI, 190 pages. 1973. DM 20,–

Vol. 310: B. Iversen, Generic Local Structure of the Morphisms in Commutative Algebra. IV, 108 pages. 1973. DM 18,–

Vol. 311: Conference on Commutative Algebra. Edited by J. W. Brewer and E. A. Rutter. VII, 251 pages. 1973. DM 24,–

Vol. 312: Symposium on Ordinary Differential Equations. Edited by W. A. Harris, Jr. and Y. Sibuya. VIII, 204 pages. 1973. DM 22,–

Vol. 313: K. Jörgens and J. Weidmann, Spectral Properties of Hamiltonian Operators. III, 140 pages. 1973. DM 18,–

Vol. 314: M. Deuring, Lectures on the Theory of Algebraic Functions of One Variable. VI, 151 pages. 1973. DM 18,–

Vol. 315: K. Bichteler, Integration Theory (with Special Attention to Vector Measures). VI, 357 pages. 1973. DM 29,–

Vol. 316: Symposium on Non-Well-Posed Problems and Logarithmic Convexity. Edited by R. J. Knops. V, 176 pages. 1973. DM 20,–

Vol. 317: Séminaire Bourbaki – vol. 1971/72. Exposés 400–417. IV, 361 pages. 1973. DM 29,–

Vol. 318: Recent Advances in Topological Dynamics. Edited by A. Beck. VIII, 285 pages. 1973. DM 27,–

Vol. 319: Conference on Group Theory. Edited by R. W. Gatterdam and K. W. Weston. V, 188 pages. 1973. DM 20,–

Vol. 320: Modular Functions of One Variable I. Edited by W. Kuyk. V, 195 pages. 1973. DM 20,–

Vol. 321: Séminaire de Probabilités VII. Edité par P. A. Meyer. VI, 322 pages. 1973. DM 29,–

Vol. 322: Nonlinear Problems in the Physical Sciences and Biology. Edited by I. Stakgold, D. D. Joseph and D. H. Sattinger. VIII, 357 pages. 1973. DM 29,–

Vol. 323: J. L. Lions, Perturbations Singulières dans les Problèmes aux Limites et en Contrôle Optimal. XII, 645 pages. 1973. DM 46,–

Vol. 324: K. Kreith, Oscillation Theory. VI, 109 pages. 1973. DM 18,–

Vol. 325: C.-C. Chou, La Transformation de Fourier Complexe et L'Equation de Convolution. IX, 137 pages. 1973. DM 18,–

Vol. 326: A. Robert, Elliptic Curves. VIII, 264 pages. 1973. DM 24,–

Vol. 327: E. Matlis, One-Dimensional Cohen-Macaulay Rings. XII, 157 pages. 1973. DM 20,–

Vol. 328: J. R. Büchi and D. Siefkes, The Monadic Second Order Theory of All Countable Ordinals. VI, 217 pages. 1973. DM 22,–

Vol. 329: W. Trebels, Multipliers for (C, α)-Bounded Fourier Expansions in Banach Spaces and Approximation Theory. VII, 103 pages. 1973. DM 18,–

Vol. 330: Proceedings of the Second Japan-USSR Symposium on Probability Theory. Edited by G. Maruyama and Yu. V. Prokhorov. VI, 550 pages. 1973. DM 40,–

Vol. 331: Summer School on Topological Vector Spaces. Edited by L. Waelbroeck. VI, 226 pages. 1973. DM 22,–

Vol. 332: Séminaire Pierre Lelong (Analyse) Année 1971-1972. V, 131 pages. 1973. DM 18,–

Vol. 333: Numerische, insbesondere approximationstheoretische Behandlung von Funktionalgleichungen. Herausgegeben von R. Ansorge und W. Törnig. VI, 296 Seiten. 1973. DM 27,–

Vol. 334: F. Schweiger, The Metrical Theory of Jacobi-Perron Algorithm. V, 111 pages. 1973. DM 18,–

Vol. 335: H. Huck, R. Roitzsch, U. Simon, W. Vortisch, R. Walden, B. Wegner und W. Wendland, Beweismethoden der Differentialgeometrie im Großen. IX, 159 Seiten. 1973. DM 20,–

Vol. 336: L'Analyse Harmonique dans le Domaine Complexe. Edité par E. J. Akutowicz. VIII, 169 pages. 1973. DM 20,–

Vol. 337: Cambridge Summer School in Mathematical Logic. Edited by A. R. D. Mathias and H. Rogers. IX, 660 pages. 1973. DM 46,–

Vol. 338: J. Lindenstrauss and L. Tzafriri, Classical Banach Spaces. IX, 243 pages. 1973. DM 24,–

Vol. 339: G. Kempf, F. Knudsen, D. Mumford and B. Saint-Donat, Toroidal Embeddings I. VIII, 209 pages. 1973. DM 22,–

Vol. 340: Groupes de Monodromie en Géométrie Algébrique. (SGA 7 II). Par P. Deligne et N. Katz. X, 438 pages. 1973. DM 44,–

Vol. 341: Algebraic K-Theory I, Higher K-Theories. Edited by H. Bass. XV, 335 pages. 1973. DM 29,–

Vol. 342: Algebraic K-Theory II, "Classical" Algebraic K-Theory, and Connections with Arithmetic. Edited by H. Bass. XV, 527 pages. 1973. DM 40,-

Vol. 343: Algebraic K-Theory III, Hermitian K-Theory and Geometric Applications. Edited by H. Bass. XV, 572 pages. 1973. DM 40,-

Vol. 344: A. S. Troelstra (Editor), Metamathematical Investigation of Intuitionistic Arithmetic and Analysis. XVII, 485 pages. 1973. DM 38,-

Vol. 345: Proceedings of a Conference on Operator Theory. Edited by P. A. Fillmore. VI, 228 pages. 1973. DM 22,-

Vol. 346: Fučík et al., Spectral Analysis of Nonlinear Operators. II, 287 pages. 1973. DM 26,-

Vol. 347: J. M. Boardman and R. M. Vogt, Homotopy Invariant Algebraic Structures on Topological Spaces. X, 257 pages. 1973. DM 24,-

Vol. 348: A. M. Mathai and R. K. Saxena, Generalized Hypergeometric Functions with Applications in Statistics and Physical Sciences. VII, 314 pages. 1973. DM 26,-

Vol. 349: Modular Functions of One Variable II. Edited by W. Kuyk and P. Deligne. V, 598 pages. 1973. DM 38,-

Vol. 350: Modular Functions of One Variable III. Edited by W. Kuyk and J.-P. Serre. V, 350 pages. 1973. DM 26,-

Vol. 351: H. Tachikawa, Quasi-Frobenius Rings and Generalizations. XI, 172 pages. 1973. DM 20,-

Vol. 352: J. D. Fay, Theta Functions on Riemann Surfaces. V, 137 pages. 1973. DM 18,-

Vol. 353: Proceedings of the Conference on Orders, Group Rings and Related Topics. Organized by J. S. Hsia, M. L. Madan and T. G. Ralley. X, 224 pages. 1973. DM 22,-

Vol. 354: K. J. Devlin, Aspects of Constructibility. XII, 240 pages. 1973. DM 24,-

Vol. 355: M. Sion, A Theory of Semigroup Valued Measures. V, 140 pages. 1973. DM 18,-

Vol. 356: W. L. J. van der Kallen, Infinitesimally Central Extensions of Chevalley Groups. VII, 147 pages. 1973. DM 18,-

Vol. 357: W. Borho, P. Gabriel und R. Rentschler, Primideale in Einhüllenden auflösbarer Lie-Algebren. V, 182 Seiten. 1973. DM 20,-

Vol. 358: F. L. Williams, Tensor Products of Principal Series Representations. VI, 132 pages. 1973. DM 18,-

Vol. 359: U. Stammbach, Homology in Group Theory. VIII, 183 pages. 1973. DM 20,-

Vol. 360: W. J. Padgett and R. L. Taylor, Laws of Large Numbers for Normed Linear Spaces and Certain Fréchet Spaces. VI, 111 pages. 1973. DM 18,-

Vol. 361: J. W. Schutz, Foundations of Special Relativity: Kinematic Axioms for Minkowski Space Time. XX, 314 pages. 1973. DM 26,-

Vol. 362: Proceedings of the Conference on Numerical Solution of Ordinary Differential Equations. Edited by D. Bettis. VIII, 490 pages. 1974. DM 34,-

Vol. 363: Conference on the Numerical Solution of Differential Equations. Edited by G. A. Watson. IX, 221 pages. 1974. DM 20,-

Vol. 364: Proceedings on Infinite Dimensional Holomorphy. Edited by T. L. Hayden and T. J. Suffridge. VII, 212 pages. 1974. DM 20,-

Vol. 365: R. P. Gilbert, Constructive Methods for Elliptic Equations. VII, 397 pages. 1974. DM 26,-

Vol. 366: R. Steinberg, Conjugacy Classes in Algebraic Groups (Notes by V. V. Deodhar). VI, 159 pages. 1974. DM 18,-

Vol. 367: K. Langmann und W. Lütkebohmert, Cousinverteilungen und Fortsetzungssätze. VI, 151 Seiten. 1974. DM 16,-

Vol. 368: R. J. Milgram, Unstable Homotopy from the Stable Point of View. V, 109 pages. 1974. DM 16,-

Vol. 369: Victoria Symposium on Nonstandard Analysis. Edited by A. Hurd and P. Loeb. XVIII, 339 pages. 1974. DM 26,-

Vol. 370: B. Mazur and W. Messing, Universal Extensions and One Dimensional Crystalline Cohomology. VII, 134 pages. 1974. DM 16,-

Vol. 371: V. Poenaru, Analyse Différentielle. V, 228 pages. 1974. DM 20,-

Vol. 372: Proceedings of the Second International Conference on the Theory of Groups 1973. Edited by M. F. Newman. VII, 740 pages. 1974. DM 48,-

Vol. 373: A. E. R. Woodcock and T. Poston, A Geometrical Study of the Elementary Catastrophes. V, 257 pages. 1974. DM 22,-

Vol. 374: S. Yamamuro, Differential Calculus in Topological Linear Spaces. IV, 179 pages. 1974. DM 18,-

Vol. 375: Topology Conference 1973. Edited by R. F. Dickman Jr. and P. Fletcher. X, 283 pages. 1974. DM 24,-

Vol. 376: D. B. Osteyee and I. J. Good, Information, Weight of Evidence, the Singularity between Probability Measures and Signal Detection. XI, 156 pages. 1974. DM 16,-

Vol. 377: A. M. Fink, Almost Periodic Differential Equations. VIII, 336 pages. 1974. DM 26,-

Vol. 378: TOPO 72 - General Topology and its Applications. Proceedings 1972. Edited by R. Alò, R. W. Heath and J. Nagata. XIV, 651 pages. 1974. DM 50,-

Vol. 379: A. Badrikian et S. Chevet, Mesures Cylindriques, Espaces de Wiener et Fonctions Aléatoires Gaussiennes. X, 383 pages. 1974. DM 32,-

Vol. 380: M. Petrich, Rings and Semigroups. VIII, 182 pages. 1974. DM 18,-

Vol. 381: Séminaire de Probabilités VIII. Edité par P. A. Meyer. IX, 354 pages. 1974. DM 32,-

Vol. 382: J. H. van Lint, Combinatorial Theory Seminar Eindhoven University of Technology. VI, 131 pages. 1974. DM 18,-

Vol. 383: Séminaire Bourbaki - vol. 1972/73. Exposés 418-435 IV, 334 pages. 1974. DM 30,-

Vol. 384: Functional Analysis and Applications, Proceedings 1972. Edited by L. Nachbin. V, 270 pages. 1974. DM 22,-

Vol. 385: J. Douglas Jr. and T. Dupont, Collocation Methods for Parabolic Equations in a Single Space Variable (Based on C^1-Piecewise-Polynomial Spaces). V, 147 pages. 1974. DM 16,-

Vol. 386: J. Tits, Buildings of Spherical Type and Finite BN-Pairs. IX, 299 pages. 1974. DM 24,-

Vol. 387: C. P. Bruter, Eléments de la Théorie des Matroïdes. V, 138 pages. 1974. DM 18,-

Vol. 388: R. L. Lipsman, Group Representations. X, 166 pages. 1974. DM 20,-

Vol. 389: M.-A. Knus et M. Ojanguren, Théorie de la Descente et Algèbres d' Azumaya. IV, 163 pages. 1974. DM 20,-

Vol. 390: P. A. Meyer, P. Priouret et F. Spitzer, Ecole d'Eté de Probabilités de Saint-Flour III - 1973. Edité par A. Badrikian et P.-L. Hennequin. VIII, 189 pages. 1974. DM 20,-

Vol. 391: J. Gray, Formal Category Theory: Adjointness for 2-Categories. XII, 282 pages. 1974. DM 24,-

Vol. 392: Géométrie Différentielle, Colloque, Santiago de Compostela, Espagne 1972. Edité par E. Vidal. VI, 225 pages. 1974. DM 20,-

Vol. 393: G. Wassermann, Stability of Unfoldings. IX, 164 pages. 1974. DM 20,-

Vol. 394: W. M. Patterson 3rd, Iterative Methods for the Solution of a Linear Operator Equation in Hilbert Space - A Survey. III, 183 pages. 1974. DM 20,-

Vol. 395: Numerische Behandlung nichtlinearer Integrodifferential- und Differentialgleichungen. Tagung 1973. Herausgegeben von R. Ansorge und W. Törnig. VII, 313 Seiten. 1974. DM 28,-

Vol. 396: K. H. Hofmann, M. Mislove and A. Stralka, The Pontryagin Duality of Compact O-Dimensional Semilattices and its Applications. XVI, 122 pages. 1974. DM 18,-

Vol. 397: T. Yamada, The Schur Subgroup of the Brauer Group. V, 159 pages. 1974. DM 18,-

Vol. 398: Théories de l'Information, Actes des Rencontres de Marseille-Luminy, 1973. Edité par J. Kampé de Fériet et C. Picard. XII, 201 pages. 1974. DM 23,-